Radiation
Protection

This book is due for return on or before the last date shown below.

An Introduction to

Radiation

Protection

THIRD EDITION

Alan Martin
Associated Nuclear Services, Epsom

and

Samuel A. Harbison
Nuclear Installations Inspectorate
Health and Safety Executive, London

London   New York

CHAPMAN AND HALL

First published in 1972 by
Chapman and Hall Ltd
11 New Fetter Lane, London EC4P 4EE
Published in the USA by
Chapman and Hall
29 West 35th Street, New York NY 10001
Second edition 1979
Reprinted 1982
Third edition 1986

© *1972, 1979, 1986 Alan Martin and Samuel A. Harbison*

Printed in Great Britain by J. W. Arrowsmith Ltd, Bristol

ISBN 0 412 27800 6 (Hardback)
ISBN 0 412 27810 3 (Science Paperback)

British Library Cataloguing in Publication Data

Martin, Alan
   An introduction to radiation protection.—3rd ed.
   1. Ionizing radiation—Safety measures
   I. Title   II. Harbison, Samuel A.
   363.1'79   RA1231.R2

   ISBN 0-412-27800-6
   ISBN 0-412-27810-3 Pbk

Library of Congress Cataloging in Publication Data

Martin, Alan
   An introduction to radiation protection.
   Bibliography: p.
   Includes index.
   1. Ionizing radiation—Safety measures.
   2. Ionizing radiation—Toxicology. I. Harbison, Samuel A.
   II. Title. [DNLM: 1. Radiation Protection.
   WN 650 M379i]
   RA569.M37   1986   604.7   86-2315
ISBN 0-412-27800-6
ISBN 0-412-27810-3 (Science paperback: pbk.)

Contents

Preface

An Introduction to Radiation Protection is a comprehensive account of radiation hazards and their control. The presentation, which assumes no previous knowledge of the subject, is based on methods which the authors found, over a number of years of teaching radiation protection at all levels, to be most easily understood by students. It is hoped that the book will meet the requirements of a wide range of readers who are involved, either directly or indirectly, with ionizing radiation, including doctors, dentists, research workers, nuclear plant designers and operators. In particular we believe that the work is suitable for the health physics monitors and technicians who are concerned with the day-to-day control of radiation hazards in nuclear power stations, research establishments, hospitals and in industry. In the UK, the generally accepted standard of training in this type of work is that set by the City and Guilds of London Institute courses in Radiation Safety Practice which are held in various centres. The chapters of the book dealing with the general aspects of health physics are aimed at this standard. Later chapters dealing with particular aspects of the subject are more detailed so that, for example, a health physics monitor in a nuclear power station or a technician in a hospital can get a deeper understanding of the problems in his or her own area.

When discussing legislation and regulations relating to radiation protection, we have illustrated the position by reference to the requirements in the UK. Otherwise the principles and practice set out in the book are internationally applicable.

In the second edition we adopted the SI units but the old units were defined and used in parentheses throughout the book. After careful consideration, we have continued this dual approach in the third edition since SI units are not yet universally in use.

Every attempt has been made to avoid detailed mathematical treatment but it has been necessary, in some areas, to use some simple mathematics. This includes squares, square roots, exponentials, logarithms and the plotting of graphs on logarithmic scales. Where a mathematical treatment is used we have tried to present it in such a way that, if the mathematics is not fully understood, it does not preclude an understanding of the chapter in general.

As far as possible each chapter is self-contained so that the reader can find all the information on a particular aspect without having to search through several chapters. The early chapters deal with basic physical principles, the nature of the hazard arising from the interaction of ionizing radiation with biological systems and the levels of radiation which are regarded as acceptable. Later chapters deal with the methods of measurement and control which are applied to attain these levels. In the second half of the book there are individual chapters on the more specialized topics of nuclear reactor health physics, problems associated with X-rays and radiography, health physics in medicine, the disposal of radioactive waste and radiological emergencies. Chapters are also presented on legislation and on the organization of health physics. Each chapter is followed by a summary in note form, in which the major points are reiterated. In addition, a number of revision questions requiring both descriptive and numerical answers are provided for the majority of chapters.

We have been greatly assisted in the preparation of this and earlier editions by detailed criticisms and suggestions by Mr H. M. Carruthers, Mr E. Greenslade, Dr R. J. Ott, Miss B. E. Stern, Dr D. Sykes and Dr J. Vennart. In particular we are grateful to Surgeon Captain Michael Hatfield RN for a critical review of the draft manuscript and many helpful and constructive suggestions. The opinions and conclusions expressed in the book are those of the authors and have no official sanction.

Finally, we are especially grateful to our wives Linda and Joyce for their patience and encouragement during the writing of the book.

London
1986

AM
SAH

1

The structure of matter

1.1 INTRODUCTION

Matter is the name given to the materials of which the Universe is composed. It exists in three physical forms: solid, liquid and gas. All matter is found to consist of a number of simple substances called elements.

An *element* is a substance which cannot be broken down by ordinary chemical processes into simpler substances. There are 92 naturally occurring elements, of which examples are carbon, oxygen, iron and lead. Another dozen or so have been produced artificially in recent years, and the best known of these is plutonium. In nature, elements are usually chemically linked to other elements in the form of compounds.

A *compound* consists of two or more elements chemically linked in definite proportions, e.g. water, H_2O, which consists of two atoms of hydrogen and one atom of oxygen.

1.2 THE ATOM

Consider an amount of some element and subject it to repeated subdivisions. Using ordinary optical instruments a stage will eventually be reached when the fragments cease to be visible. Supposing, however, that suitable tools and viewing apparatus were available, would it be possible to repeat the divisions of the original element indefinitely or would a stage be reached where the matter can no longer be subdivided?

More than 2000 years ago, the Greek philosophers considered this question. They did not have any viewing apparatus with which to observe the actual division of an element into vanishingly-small fragments. All they could do was to think about the problem in a logical manner. From this philosophical approach some of them decided that eventually a limit must be reached. They called the individual particles of matter, which could not be further subdivided, *atoms*. It was also postulated by some of the philosophers that all substances consist of these same atoms, different arrangements of the constituent atoms giving the different properties of the substances and the density being determined by how tightly the atoms are packed.

Early in the nineteenth century, an atomic theory with a scientific basis was advanced which confirmed many of the views held by the ancient philosophers. This was the atomic theory of Dalton which was able to explain the well established but little understood chemical laws. Modern theory has diverged somewhat from Dalton's but he did establish the principle that matter consists of atoms, each element having its own characteristic atom.

1.3   THE STRUCTURE OF THE ATOM

It is now known that atoms are not solid, indivisible objects as the Greek philosophers believed but are composed of even smaller particles. These particles, from which all atoms are constructed, are called *protons, neutrons* and *electrons*.

The *proton* (p) carries a positive electrical charge of magnitude one unit on the nuclear scale, and a mass of approximately one *atomic mass unit* (u).

The *electron* (e^-) has a negative electrical charge of the same magnitude as the proton's positive charge. It has a mass of 1/1840 u, which for most purposes is neglected in considering the mass of the atom.

The *neutron* (n) is often regarded as a close combination of a proton and an electron. It is electrically *neutral* and, neglecting the mass of the electron, has a mass of approximately one atomic mass unit. In the text and in illustrations the neutron will generally be treated as a fundamental particle, in common with normal usage.

The neutrons and protons of an atom form a central core or *nucleus* around which the electrons rotate in various orbits. It is found that the orbit closest to the nucleus can contain a maximum of 2 electrons, while the second can have up to 8 electrons, and so on for the outer orbits. The inner orbit is called the *K orbit* (or K shell), the second orbit is called the *L shell*, the third the *M shell*, and so on. The maximum numbers of electrons in the K, L, M, N shells are 2, 8, 18, 32

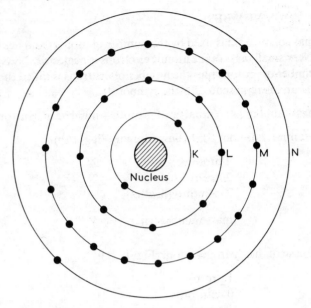

Fig. 1.1 The atomic system of zinc

respectively. For example, the atomic system of zinc, illustrated in Fig. 1.1 has 30 electrons arranged in 4 shells.

Each atom normally has the same number of protons as electrons. This means that the total positive charge on the nucleus is equal to the total negative charge due to the atomic electrons and so the atom is normally electrically neutral. Two simple atoms are illustrated in Fig. 1.2. The hydrogen atom is the simplest of all atoms and is the only atom which does not contain neutrons.

Fig. 1.2 Atomic systems of hydrogen and helium

1.4   MASS NUMBER

The mass of an atom is fixed by the number of protons and neutrons, if the very small mass of the atomic electrons is neglected. The sum of the number of protons plus the number of neutrons is called the *mass number* and is represented by the symbol A:

mass number (A) = number of protons + number of neutrons

For example, consider the helium atom in Fig. 1.2(b)

2 protons	2
2 neutrons	2
2 electrons (neglect)	–
mass number (A)	= 4

In the case of the hydrogen atom (Fig. 1.2(a))

1 proton	1
0 neutrons	0
1 electron (neglect)	–
mass number (A)	= 1

1.5   ATOMIC NUMBER

The number of protons in an atom is called the *atomic number*, represented by the symbol Z:

Atomic number (Z) = number of protons
eg
      helium has 2 protons, $Z = 2$
      hydrogen has 1 proton, $Z = 1$

The chemical symbol for helium is He and helium atoms of the type mentioned above would be designated $^4_2$He, and the hydrogen atoms (symbol H), $^1_1$H.

   The number of protons, that is the atomic number, determines the chemical properties of the atom and so defines the element.

Thus

    all atoms with an atomic number of 1 are hydrogen atoms
    all atoms with an atomic number of 2 are helium atoms

all atoms with an atomic number of 3 are lithium atoms
all atoms with an atomic number of 4 are beryllium atoms
all atoms with an atomic number of 5 are boron atoms
all atoms with an atomic number of 6 are carbon atoms . . . , etc.,

up to the heaviest naturally-occurring element, uranium, which has an atomic number of 92. About a dozen or so elements of higher atomic number have been artificially produced in recent years. They are all unstable and can only be made under special conditions which are not found naturally on Earth.

1.6 ISOTOPES

Although all the atoms of a particular element contain the same number of protons, they may occur with different numbers of neutrons. This means that an element can have several types of atom. For example, the element phosphorus (P) has an atomic number of 15 (i.e. each atom contains 15 protons), but it can occur with different numbers of neutrons:

$^{28}_{15}P$	15 protons, 13 neutrons	$(Z=15, A=28)$
$^{29}_{15}P$	15 protons, 14 neutrons	$(Z=15, A=29)$
$^{30}_{15}P$	15 protons, 15 neutrons	$(Z=15, A=30)$
$^{31}_{15}P$	15 protons, 16 neutrons	$(Z=15, A=31)$
$^{32}_{15}P$	15 protons, 17 neutrons	$(Z=15, A=32)$
$^{33}_{15}P$	15 protons, 18 neutrons	$(Z=15, A=33)$
$^{34}_{15}P$	15 protons, 19 neutrons	$(Z=15, A=34)$

These different forms are called *isotopes* of the element. Thus, for example, $^{32}_{15}P$ is an isotope of phosphorus. The three isotopes of helium $(Z=2)$ are illustrated in Fig. 1.3. These isotopes are normally referred to as helium-3, helium-4 and helium-5, respectively (^{3}He, ^{4}He, ^{5}He).

It is important to note that all the isotopes of a given element are *chemically* identical, since the chemical properties are determined by the atomic number of the element.

Most elements occur naturally as a mixture of isotopes and other isotopes may be produced by bombarding a naturally occurring isotope with nuclear particles, for example by neutrons in a nuclear reactor. These artificially produced isotopes are unstable and will

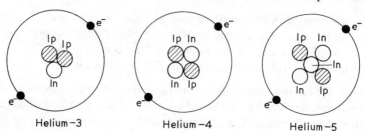

Fig. 1.3   The three isotopes of helium

eventually disintegrate with the emission of a secondary particle (see Chapter 2).

Apart from the few lightest elements the number of neutrons exceeds the number of protons in an atom. The difference becomes greater as Z increases, e.g.

$$^4_2\text{He} \qquad \text{2 protons} + \text{2 neutrons}$$

$$^{31}_{15}\text{P} \qquad \text{15 protons} + \text{16 neutrons}$$

$$^{65}_{30}\text{Zn} \qquad \text{30 protons} + \text{35 neutrons}$$

$$^{238}_{92}\text{U} \qquad \text{92 protons} + \text{146 neutrons}$$

Data on the known isotopes of all the elements, both naturally occurring and aritificially produced, have been arranged systematically in a table known as the *chart of the nuclides* and this will be discussed in more detail in Chapter 2. The term *nuclide* means any isotope of any element.

1.7   ANCIENT AND MODERN THEORIES

It will now be seen that the ancient Greek philosophers were remarkably close to the truth in their theory that all substances are constituted from the same basic particles. However, instead of being different arrangements of only one type of particle, different substances appear to result from various combinations of three 'fundamental' particles: protons, neutrons and electrons. Nuclear physicists are still finding other apparently more fundamental particles, so that the Ancient Greeks may yet prove to have been correct.

SUMMARY OF CHAPTER

Element: material whose atoms all have the same number of protons.
u: atomic mass unit.
Proton: fundamental particle, mass 1 u, charge + 1 unit.
Electron: fundamental particle, mass 1/1840 u, charge − 1 unit.
Neutron: close combination of proton and electron, mass 1 u, electrically neutral.
Atom: central nucleus of protons and neutrons, around which electrons rotate in orbits.
Mass number (*A*): number of protons + number of neutrons.
Atomic number (*Z*): number of protons.
Isotope: one of several nuclides with the same atomic number.
Nuclide: a nuclear species.

REVISION QUESTIONS

1. Using a billiard ball technique, draw an atom of each of the following nuclides:

$$^4_2\text{He}, \quad ^6_3\text{Li}, \quad ^7_3\text{Li}, \quad ^8_4\text{Be}$$

2. The nuclide cobalt 60 may be written $^{60}_{27}\text{Co}$. How many protons, electrons and neutrons are there in this type of atom?

3. What are the masses and charges on the atomic scale of protons, electrons and neutrons?

4. Which atomic property determines the chemical behaviour of an element?

5. Explain what is meant by the term *isotope*. Give some examples.

2

Radioactivity and radiation

2.1 INTRODUCTION

It is found that a few naturally occurring substances consist of atoms which are unstable – that is they undergo spontaneous transformation into more stable product atoms. Such substances are said to be *radioactive* and the transformation process is known as *radioactive decay*. Radioactive decay is usually accompanied by the emission of radiation in the form of charged particles and gamma rays.

The fact that some elements are naturally radioactive was first realized by Becquerel in 1896. He observed blackening of photographic emulsions in the vicinity of a uranium compound. This was subsequently attributed to radiation being emitted by the uranium. In the following ten years the outstanding experimental work of Rutherford and Soddy, M. and Mme Curie and others established the fact that certain nuclei are not completely stable. These unstable nuclei emit radiations of three main types, called *alpha, beta* and *gamma* radiation.

2.2 ALPHA, BETA AND GAMMA RADIATION

Alpha radiation (α) was shown by Rutherford and Royds to consist of helium nuclei which themselves consist of two protons and two neutrons. These four particles are bound together so tightly that the alpha particle behaves in many situations as if it were a fundamental particle. An α-particle has a mass of 4 u and carries two units of positive charge.

Beta radiation (β) consists of high speed electrons which originate in the nucleus. These 'nuclear electrons' have identical properties to the atomic electrons, that is they have a mass of 1/1840 u and carry one unit of negative charge. Another type of beta radiation was discovered by C. D. Anderson in 1932. This consists of particles of the same mass as the electron but having one unit of positive charge, and is known as *positron* radiation. Although less important from a radiation protection viewpoint than negative β-particles a knowledge of positrons is necessary in order to understand certain radioactive decay mechanisms. Beta radiation is signified β^- (electrons) or β^+ (positrons). In everyday use the term beta radiation normally refers to the negative type, β^-.

Gamma radiation (γ) belongs to a class known as electromagnetic radiation. This type of radiation consists of *quanta* or packets of energy transmitted in the form of a wave motion. Other well-known members of this class of radiation are radio waves and visible light. The amount of energy in each quantum is related to the wavelength of the radiation. The energy is inversely proportional to the wavelength, i.e. $E \propto 1/\lambda$ where E is the energy of the *quantum* or *photon* of electromagnetic radiation and λ is its wavelength.

The wavelength of electromagnetic radiation varies over a very wide range, as illustrated in Table 2.1.

Table 2.1

Type of radiation	Wavelength, λ metres
Radio waves, BBC Radio 4	1500
Radio waves, VHF Band 2	3
Visible light	10^{-6} to 10^{-7}
X-rays, 50 keV energy	$2 \cdot 5 \times 10^{-11}$
γ-rays, 1 MeV energy	$1 \cdot 2 \times 10^{-12}$

All electromagnetic radiations travel through free space with the same velocity of 3×10^8 m/s (186 000 miles/s). Their velocity decreases in dense media but for air the decrease is negligible.

Another class of electromagnetic radiation which is in most respects identical to γ-radiation is known as X-radiation. The essential difference between the two types of radiation lies in their origin. Whereas γ-rays result from changes in the nucleus. X-rays are emitted when atomic electrons undergo a change in orbit.

2.3 THE ELECTRONVOLT

Radiation energy is expressed in *electronvolts* (eV). One electronvolt is the energy gained by an electron in passing through an electrical potential of one volt.

For example: in the cathode ray tube of a television receiver, electrons are accelerated from the electron gun to the screen through an electrical potential of about 10 000 volts. The electrons therefore have an energy of 10 000 eV when they strike the screen.

The electronvolt is a very small unit so radiation energies are usually expressed in *kilo* (1000) or *mega* (1 000 000) electronvolts:

One kiloelectronvolt = 1 keV = 1000 eV
One megaelectronvolt = 1 MeV = 1000 keV = 1 000 000 eV

Even if the radiation being considered is not beta (electron) radiation, it is still possible to use the electronvolt as a unit of energy. The energy of a particle depends on its mass and speed, e.g. a particle of mass m travelling with velocity v, much smaller than the velocity of light, has kinetic energy (E_K)

$$E_K = \tfrac{1}{2}mv^2$$

(a correction is necessary for particles having velocities approaching the velocity of light). A small particle such as an electron requires a much higher velocity than, say, an α-particle to have the same kinetic energy.

In the case of electromagnetic radiation, the energy is inversely proportional to the wavelength of the radiation. Thus radiations with short wavelengths have higher energies than radiations with longer wavelengths.

2.4 THE MECHANISM OF RADIOACTIVE DECAY

The nuclei of the heavier elements found in nature are so large that they are slightly unstable. For example, the isotope uranium-238 has 92 protons and 146 neutrons. In order to achieve greater stability the nucleus may emit an alpha particle, so reducing its numbers of protons and neutrons to 90 and 144, respectively. This means that the nucleus now has an atomic number (Z) of 90 instead of 92 and so is no longer a uranium nucleus. It is now an isotope of the element thorium

(Th) with atomic number 90 and mass number 234, namely thorium-234. This decay process may be represented as follows:

$$^{238}_{92}\text{U} \longrightarrow {}^{4}_{2}\alpha + {}^{234}_{90}\text{Th}$$

or, more commonly,

$$^{238}_{92}\text{U} \xrightarrow{\alpha} {}^{234}_{90}\text{Th}$$

Another example of this process is the decay of polonium-218 (^{218}Po) by alpha emission to lead-214 (^{214}Pb):

$$^{218}_{84}\text{Po} \xrightarrow{\alpha} {}^{214}_{82}\text{Pb}$$

It was pointed out in Chapter 1 that there are more neutrons than protons in heavy nuclei. Alpha emission reduces the number of each by 2 but the proportionate reduction is considerably less for neutrons than for protons. In the ^{238}U decay process, the number of protons is reduced by 2 out of 92 whereas the number of neutrons is reduced by 2 out of 146 which is significantly less. The effect of alpha emission is therefore to produce neutron-rich nuclei which are still unstable. The nucleus does not simply eject a neutron (or neutrons) to correct this instability. Instead, one of the neutrons in the nucleus changes into a proton by emitting a beta particle, i.e. a high speed electron:

$$^{1}_{0}\text{n} \longrightarrow {}^{1}_{1}\text{p} + \beta^{-}$$

This phenomenon is known as *beta emission*. In the case of ^{234}Th, formed by the α-decay of ^{238}U, the nucleus further decays by β-emission to protactinium-234 (^{234}Pa):

$$^{234}_{90}\text{Th} \longrightarrow {}^{234}_{91}\text{Pa} + \beta^{-}$$

or

$$^{234}_{90}\text{Th} \xrightarrow{\beta^{-}} {}^{234}_{91}\text{Pa}$$

Considering again polonium-218, the complete decay is:

$$^{218}_{84}\text{Po} \xrightarrow{\alpha} {}^{214}_{82}\text{Pb} \xrightarrow{\beta^{-}} {}^{214}_{83}\text{Bi}$$

The resulting atom is bismuth-214, which is also unstable and so further α- and β-decay processes occur until a stable atom is produced.

Electrons emitted during β-decay have a continuous distribution in energy, ranging from zero to a maximum energy E_{max} which is characteristic of the particular nucleus. The most probable beta energy is about $1/3E_{max}$ (see Fig. 2.1).

In most cases, after the emission of an α- or a β-particle, the nucleus re-arranges itself slightly, releasing energy in the form of *gamma emission*.

Two other decay processes should also be mentioned, namely *positron emission* and *electron capture*. In positron emission, a proton in the nucleus ejects a positive electron (β^+) and so becomes a neutron:

$$_1^1 p \longrightarrow {}_0^1 n + \beta^+$$

For example:

$$_{11}^{22}\text{Na} \xrightarrow{\beta-} {}_{10}^{22}\text{Ne}$$

Electron capture is a process in which an electron from an inner orbit is captured by the nucleus resulting in the conversion of a proton to a neutron:

$$_1^1 p + e^- \longrightarrow {}_0^1 n$$

A re-arrangement of atomic electrons then causes the emission of X-rays.

Fig. 2.1   Typical beta spectrum

2.5   NATURAL RADIOACTIVE SERIES

Apart from ^{22}Na, the above examples of radioactive decay are all naturally-occurring radioactive substances and belong to the so-called *natural radioactive series*. There are three natural radioactive series, called the thorium, uranium–radium and actinium series (see Table 2.2). Also included in this table is the neptunium series which does not occur in nature because the half-life of its longest lived member is only 2.2×10^6 years which is much smaller than the age of the Universe (3×10^9 years). These four series are called the heavy disintegration series.

Table 2.2   Heavy disintegration series

Series name	Final stable nucleus	Longest-lived member
Thorium	^{208}Pb	^{232}Th ($T_{1/2} = 1.39 \times 10^{10}$ year)
Uranium–radium	^{206}Pb	^{238}U ($T_{1/2} = 4.50 \times 10^{9}$ year)
Actinium	^{207}Pb	^{235}U ($T_{1/2} = 8.52 \times 10^{8}$ year)
Neptunium	^{209}Bi	^{237}Np ($T_{1/2} = 2.20 \times 10^{6}$ year)

2.6   INDUCED RADIOACTIVITY

Lighter elements may be made radioactive by bombarding them with nuclear particles. One example of this is the bombardment of stable nuclei of an element by neutrons in a nuclear reactor. A neutron may be captured by a nucleus, with the emission of a gamma photon. This is known as a neutron, gamma (n, γ) reaction. The resulting atom is usually unstable because of the excess neutron and will eventually decay by β-emission.

Thus if the stable isotope cobalt-59 is bombarded or irradiated with neutrons, atoms of the radioactive isotope cobalt-60 are produced. These atoms will eventually undergo β-decay and become atoms of the stable isotope nickel-60. This process is written as:

$$^{59}_{27}\text{Co(n, } \gamma)^{60}_{27}\text{Co} \xrightarrow{\beta-} {}^{60}_{28}\text{Ni}$$

There are various other activation and decay processes which will be discussed later.

2.7   THE UNIT OF RADIOACTIVITY

The decay of a radioactive sample is statistical in nature and it is impossible to predict when any particular atom will distintegrate. The result of this random behaviour of any particular atom is that the radioactive decay law is exponential in nature, and is expressed mathematically as:

$$N = N_0 e^{-\lambda t}$$

where N_0 is the number of nuclei present initially, N is the number of nuclei present at time t and λ is the radioactive decay constant.

The *half-life* $(T_{1/2})$ of a radioactive species is the time required for one half of the nuclei in a sample to decay. It is obtained by putting $N = N_0/2$ in the above equation:

$$N_0/2 = N_0 e^{-\lambda T_{1/2}}$$

Dividing across by N_0 and taking logs

$$\log_e(\tfrac{1}{2}) = -\lambda T_{1/2}$$

Now

$$\log_e(\tfrac{1}{2}) = -\log_e(2)$$

and so

$$T_{1/2} = \frac{\log_e 2}{\lambda} = \frac{0 \cdot 693}{\lambda}$$

Since the distintegration rate, or *activity* of the sample is proportional to the number of unstable nuclei, this also varies exponentially with time, namely

$$A = A_0 e^{-\lambda t}$$

This relationship is illustrated in Fig. 2.2 which shows the variation of sample activity with time. In one half-life the activity decays to $1/2 \, A_0$, in two half-lives to $1/4 \, A_0$, and so on. The half-life of a particular radioactive isotope is constant and its measurement assists in the identification of radioactive samples of unknown composition. This method can only be applied to isotopes whose disintegration rates change appreciably over reasonable counting periods. At the other end of the scale, the isotope must have a long enough half-life to allow some measurements to be made before it all disintegrates. To

Fig. 2.2 Variation of activity with time

determine extremely short and extremely long half-lives, more elaborate means must be used. Half-lives range from about 10^{-14} years (^{212}Po) to about 10^{17} years (^{209}Bi) which represents a factor of 10^{31}.

Until quite recently the unit of radioactivity was the *curie* (Ci), along with its various submultiples. The curie was originally related to the activity of one gram of radium but the definition was later standardized as 3.7×10^{10} nuclear distintegrations (dis) per second, which is almost the same:

1 curie (Ci)	$= 3.7 \times 10^{10}$ dis/s or 2.22×10^{12} dis/min
1 millicurie (mCi)	$= 3.7 \times 10^{7}$ dis/s or 2.22×10^{9} dis/min
1 microcurie (μCi)	$= 3.7 \times 10^{4}$ dis/s or 2.22×10^{6} dis/min

A distintegration usually involves the emission of one or more charged particles (α or β). These may be accompanied, though not always, by one or more gamma emissions. Some nuclides emit only X- or gamma radiation.

The SI unit of activity is the *becquerel* (Bq) which is defined as 1 nuclear disintegration per second. Compared with the curie the becquerel is a very small unit. In practice it is often convenient to adopt the usual multiplying prefixes, for example:

$$1 \text{ becquerel (Bq)} = 1 \text{ dis/s}$$
$$1 \text{ kilobecquerel (kBq)} = 10^3 \text{ Bq} = 10^3 \text{ dis/s}$$
$$1 \text{ megabecquerel (MBq)} = 10^6 \text{ Bq} = 10^6 \text{ dis/s}$$
$$1 \text{ gigabecquerel (GBq)} = 10^9 \text{ Bq} = 10^9 \text{ dis/s}$$
$$1 \text{ terabecquerel (TBq)} = 10^{12} \text{ Bq} = 10^{12} \text{ dis/s}$$
$$1 \text{ petabecquerel (PBq)} = 10^{15} \text{ Bq} = 10^{15} \text{ dis/s}$$

For simplicity, in this text only Bq, MBq and TBq have been used. The relationship between the old unit and the SI unit is illustrated as follows:

$$1 \ \mu\text{Ci} \equiv 37\ 000 \text{ Bq} \equiv 0.037 \text{ MBq} \equiv 3.7 \times 10^{-8} \text{ TBq}$$
$$1 \text{ mCi} \equiv 37 \times 10^6 \text{ Bq} \equiv 37 \text{ MBq} \equiv 3.7 \times 10^{-5} \text{ TBq}$$
$$1 \text{ Ci} \equiv 3.7 \times 10^{10} \text{ Bq} \equiv 3.7 \times 10^4 \text{ MBq} \equiv 0.037 \text{ TBq}$$

Similarly

$$1 \text{ Bq} \equiv 2.7 \times 10^{-5} \ \mu\text{Ci} \equiv 2.7 \times 10^{-8} \text{ mCi} \equiv 2.7 \times 10^{-11} \text{ Ci}$$
$$1 \text{ MBq} \equiv 27 \ \mu\text{Ci} \equiv 0.027 \text{ mCi} \equiv 2.7 \times 10^{-5} \text{ Ci}$$
$$1 \text{ TBq} \equiv 2.7 \times 10^7 \ \mu\text{Ci} \equiv 2.7 \times 10^4 \text{ mCi} \equiv 27 \text{ Ci}$$

2.8 THE NUCLIDE CHART

A vast amount of information has been collected regarding the behaviour of stable and unstable nuclides. This information has been published in a form which summarizes the main characteristics of all known nuclides. The compilation is known as the *nuclide chart* and a portion of it is reproduced in Fig. 2.3.

In the chart, each horizontal line represents an element and the squares on that line represent the nuclides or isotopes of the element. Relevant information regarding the nuclide is printed inside the square. Stable, naturally radioactive and artificial nuclides are differentiated by the use of different colours or shading of the squares. In each case the symbol and mass number are shown as well as the natural abundance of the isotope, if it is stable. For radioactive isotopes the half-life, the mode or modes of decay, and the main energies of the emitted particles or gamma rays are shown. In the chart illustrated in Fig. 2.3, all the nuclides on the same horizontal

Fig. 2.3 A portion of the nuclide chart

line have the same atomic number while all nuclides with the same mass number lie on a 45° diagonal line, running from upper left to lower right. Many nuclide charts contain additional information which has been omitted from the sample chart shown in Fig. 2.3 for the sake of clarity.

The nuclide chart can be used to obtain rapid information on the products of various nuclear reactions. For instance, a (n, γ) reaction on sodium-23 (^{23}Na) produces sodium-24 (^{24}Na). ^{24}Na decays with a half-life of 15.0 h by emitting β^--particles of energy 1·39 MeV and γ-rays of 2·75 MeV and 1·37 MeV. The nucleus resulting from the decay of ^{24}Na is magnesium-24 (^{24}Mg) which is stable.

It is obvious that, even in the simplified form shown in Fig. 2.3, the nuclide chart is an extremely valuable source of information on the properties of both stable and unstable nuclides.

2.9   INTERACTION OF RADIATION WITH MATTER

2.9.1   Charged particles

Alpha and beta particles lose energy mainly through interactions with atomic electrons in the absorbing medium. The energy transferred to the electrons causes them either to be excited to a higher energy level (excitation) or separated entirely from the parent atom (ionization). Another important effect is that when charged particles are slowed down very rapidly they emit energy in the form of X-rays. This is known as *bremsstrahlung* (braking radiation) and is only of practical importance in the case of beta radiation.

2.9.2   X- and γ-radiations

X- and γ-radiations interact with matter through a variety of alternative mechanisms, the three most important of which are the *photoelectric effect*, *Compton scattering* and *pair-production*. In the photoelectric effect all the energy of an X- or γ-photon is transferred to an atomic electron which is ejected from its parent atom. The photon is, in this case, completely absorbed. On the other hand Compton scattering occurs when only part of the energy of the photon is transferred to an atomic electron. The photon is therefore scattered with a reduced energy.

In the intense electric field close to a charged particle, usually a nucleus, an energetic γ-photon may be converted into a positron–electron pair. This is pair-production and the two particles share the available energy.

Thus all three interactions result in the photon energy being transferred to atomic electrons which subsequently lose energy as described in Section 2.9.1.

2.9.3 Neutrons

Neutrons are uncharged and cannot cause ionization directly. As with γ-radiation, neutrons ultimately transfer their energy to charged particles. In addition a neutron may be captured by a nucleus usually resulting in γ-emission. These processes are described in greater detail in Chapter 8.

Table 2.3 summarizes the types of interactions of nuclear radiations with matter.

2.10 PENETRATING POWERS OF NUCLEAR RADIATIONS

The alpha particle is a massive particle (by nuclear standards) and travels relatively slowly through matter. It thus has a high chance of

Table 2.3 Interactions of nuclear radiations

Radiation	Process	Remarks
Alpha	Inelastic collisions with bound electrons	Leads to excitation and ionization
Beta	(i) Inelastic collisions with atomic electrons	Leads to excitation and ionization
	(ii) Slowing-down in field of nucleus	Leads to emission of bremsstrahlung
X- and γ-radiation	(i) Photoelectric effect	Photon is completely absorbed
	(ii) Compton effect	⎫ Only part of the photon
	(iii) Pair production	⎭ energy is absorbed
Neutron	(i) Elastic scattering	
	(ii) Inelastic scattering	Discussed in Chapter 8
	(iii) Capture processes	

interacting with atoms along its path and it will give up some of its energy during each of these interactions. As a consequence, alpha particles lose their energy very rapidly and only travel very short distances in dense media.

Beta particles are very much smaller than alpha particles and travel much faster. Thus they undergo fewer interactions per unit length of track and so give up their energy more slowly than alpha particles. This means that beta particles travel further in dense media than alphas.

Gamma radiation loses its energy mainly by interacting with atomic electrons. It travels very large distances in dense media and is very difficult to absorb completely.

Neutrons give up their energy through a variety of interactions, the relative importances of which are very dependent on the neutron energy. For this reason it is common practice to divide neutrons into at least three energy groups: fast, intermediate and thermal. Neutrons are very penetrating and will travel large distances in dense media.

In Table 2.4 the properties and ranges of the various nuclear radiations are summarized. The ranges are only approximate since they depend on the energy of the radiation.

Table 2.4   Properties of nuclear radiations

Radiation	Mass (u)	Charge	Range in air	Range in tissue
Alpha	4	+2	0·03 m	0·04 mm
Beta	$\frac{1}{1840}$	−1(+1 positron)	3 m	5 mm
X-, γ-radiation	0	0	V. large	Through body
Fast neutron	1	0	V. large	Through body
Thermal neutron	1	0	V. large	0·15 m

SUMMARY OF CHAPTER

Radioactive decay: transformation of an unstable substance into a more stable form, usually accompanied by the emission of charged particles and gamma rays.

Alpha radiation: helium nuclei, 2p + 2n, mass 4 u, charge + 2 units.

Beta radiation: high speed electrons which originate in the nucleus, mass 1/1840 u, charge -1 (electron) or $+1$ (positron).

Gamma radiation: electromagnetic radiation, very short wavelength, $E \propto 1/\lambda$, mass 0, charge 0.

Electronvolt: energy gained by an electron in passing through an electric potential of one volt.

$$10^6 \text{ eV} \equiv 10^3 \text{ keV} \equiv 1 \text{ MeV}$$

Natural radioactive series: consist of naturally occurring radioactive substances; the three series are thorium, uranium–radium and actinium.

Radioactive decay law: $N = N_0 e^{-\lambda t}$.

Half-life: time required for one half of the nuclei of a radioactive species to decay:

$$T_{1/2} = \frac{0.693}{\lambda}$$

Curie (Ci): former unit of radioactivity, defined as 3.7×10^{10} dis/s

$$1 \text{ Ci} \equiv 10^3 \text{ mCi} \equiv 10^6 \text{ } \mu\text{Ci}$$

Becquerel (Bq): unit of radioactivity, defined as 1 dis/s

$$1 \text{ TBq} \equiv 10^6 \text{ MBq} \equiv 10^{12} \text{ Bq}$$

Nuclide chart: compilation of data on all known nuclides.

Alpha particles lose energy in matter through excitation and ionization.

Beta particles lose energy by:

(a) excitation and ionization of atomic electrons,
(b) rapid slowing down with emission of bremsstrahlung.

γ-photons lose energy through:

(a) photoelectric effect,
(b) Compton effect,
(c) pair-production.

Neutrons lose energy through:

(a) elastic scatter,
(b) inelastic scatter,
(c) capture reactions.

REVISION QUESTIONS

1. Explain the difference between radioactivity and radiation.

2. Name the products of the following radioactive decay processes:

   (a) α-decay of uranium-238, $^{238}_{92}U$,
   (b) β^--decay of tritium, ^3_1H,
   (c) β^+-decay of copper-62, $^{62}_{29}Cu$.

3. Explain why there are only three naturally occurring radioactive series.

4. Estimate the half-life of a radioactive sample from the following series of measurements:

Time (min)	0	1	2	3	4	5	6	7	8
Activity (counts min^{-1})	—	820	605	447	330	243	180	133	98

5. Express the following activities in megabecquerels (MBq):

   (a) 5×10^6 dis/s,
   (b) 750 kBq,
   (c) 1·3 GBq,
   (d) 6×10^7 dis/min.

6. Convert the following activities to becquerels:

   (a) 1 μCi,
   (b) 5 mCi,
   (c) 2·7 Ci,
   (d) 10^6 Ci.

7. Why is an α-decay usually followed by a β^--decay?

8. Using a nuclide chart, write down the product or sequence of products which would result from a (n, γ) capture in the following nuclei:

$$^{59}_{27}Fe, \qquad ^{23}_{11}Na, \qquad ^{239}_{94}Pu$$

3

Radiation units

3.1 ABSORPTION OF ENERGY

Just as heat and light transfer energy from the Sun to the Earth and the atmosphere, so does nuclear radiation transfer energy from a source to an absorbing medium. The source of nuclear radiation may be radioactive atoms or equipment such as X-ray machines. The effect of absorbing the more familiar types of radiation, such as heat, is to raise the temperature of the absorbing medium. If this medium is the human body, or part of it, the rise in temperature is sensed and, if it becomes excessive, avoiding action can be taken, by sheltering under a sunshade (shielding), for example, or by moving further away from a fire (distance).

A dose of gamma or other nuclear radiation large enough to be lethal to a human being would increase the body temperature by less than one thousandth of one degree centigrade. The body is therefore unable to sense even very high intensities of these types of radiation. Nuclear radiation differs from heat and other types of radiation in that each particle or photon has a sufficiently high energy to cause *ionization*. The high energy is due to the very high velocity of the particles or the short wavelength of the X- and γ-radiation.

3.2 IONIZATION

Ionization is the removal of an orbital electron from an atom. Since the electron has a negative charge, the atom is consequently left positively charged. The atom and the electron, so separated, are known as an *ion pair*, that is, a positive ion (the atom) and a negative ion (the electron). The absorption of radiation in a medium results, then, in the production of ion pairs in the medium. A small amount of energy is required to cause ionization, so that in producing ion pairs, the particles or photons of radiation lose energy to the medium.

Fig. 3.1 Ionization of a helium atom by an α-particle

Figure 3.1 shows the ionization of a helium atom by an α-particle.

Normally positive and negative ions recombine to form neutral atoms and the energy originally given to the ion pair is converted into heat energy. If the absorbing medium is a gas, such as air, the ions can be prevented from recombining by applying an electric field. This is done by applying a voltage between two plates (electrodes) with a gas gap between them. Figure 3.2 shows such a system, known as an ionization chamber, in which ion pairs are being produced along the track of a beta particle. If the applied voltage is sufficiently high, negative ions produced in the volume between the electrodes are attracted to the positive electrode and positive ions to the negative electrode. The flow of ions to the respective electrodes constitutes an electrical current and since this is proportional to the intensity of radiation, ion chambers provide a means of measuring radiation. It

Fig. 3.2 Ionization chamber system

should be realized that, although only a few ion pairs are shown in the figure, in the case of beta particles, several hundreds of ion pairs are formed per centimetre of track in air, and in the case of alpha particles some tens of thousands.

In media such as water (of which the human body is largely composed) ionization can lead to breakdown of water molecules and the formation of chemical forms which are damaging to biological material. The harmful effects of radiation on the human system, which are described in Chapter 4, are largely attributable to such chemical reactions.

As already mentioned, the ionization of a gas provides a means of detecting radiation and the first widely used radiation unit, the *roentgen*, was based on the ionizing effect on air of X- and γ-radiation. This unit had several limitations and so two further units, the *rad* and the *rem*, were introduced. More recently, these two units have been replaced in the SI system by the *gray* and the *sievert*, respectively.

The *gray* and the *sievert* have been approved by the International Commission on Radiation Units and Measurements (ICRU) and used by the International Commission on Radiological Protection (ICRP). However, it is inevitable that the older units, the rad and the rem, will continue to be used for a number of years.

3.3 RADIATION ABSORBED DOSE

Absorbed dose is a measure of energy deposition in any medium by any type of ionizing radiation. The original unit of absorbed dose was the *rad* and was defined as an energy deposition of 0·01 J/kg.

In the SI system of units the unit of absorbed dose is called the *gray* (Gy) and is defined as an energy deposition of 1 J/kg. Thus:

$$1 \text{ Gy} = 1 \text{ J/kg} = 100 \text{ rad}$$

When quoting an absorbed dose, it is important to specify the absorbing medium.

3.4 DOSE EQUIVALENT

Although the quantity *absorbed dose* is a very useful physical concept, it transpires that in biological systems the same degree of damage is not necessarily produced by the same absorbed dose of different types of radiation. It is found, for example, that 0·05 Gy (5 rad) of fast neutrons can do as much biological damage as 1 Gy (100 rad) of gamma radiation. This difference in the radiobiological effectiveness must be taken into account if we wish to add doses of different radiations to obtain the total biologically effective dose. To do this we must multiply the absorbed dose of each type of radiation by a *quality factor* (Q) which reflects the ability of the particular type of radiation to cause damage. The quantity obtained when absorbed dose is multiplied by quality factor is known as the *dose equivalent*, the unit of which was originally the *rem*:

$$\text{dose equivalent (rem)} = \text{absorbed dose (rad)} \times Q$$

In the SI system of units the unit of dose equivalent is the *sievert*, abbreviated to Sv, which is related to the gray as follows:

$$\text{dose equivalent (Sv)} = \text{absorbed dose (Gy)} \times Q \times N$$

where N is a further modifying factor which might take account of such factors as absorbed dose rate and fractionation. For the present, ICRP has assigned a value of 1 to N. Since 1 Gy = 100 rad it follows that 1 Sv = 100 rem.

The value of the quality factor is found to depend on the density of ionization caused by the radiation. An alpha particle produces about 1 million ion pairs per millimetre of track in tissue whereas a beta particle produces about ten thousand per millimetre. The quality factor, Q, is assigned a value of 1 for gamma radiation and the values for other types of radiation are related to this in accordance with their ionization densities. Beta radiation causes ionization of a similar density to gamma radiation and so the quality factor is also one for beta radiation. The Q for neutrons depends on the neutron energy. For thermal neutrons Q is about 5 and for fast neutrons it is 20. For alpha and other multiply-charged particles Q is also taken as 20. The values of Q for the most commonly encountered radiations are summarized in Table 3.1.

Table 3.1   Summary of values of Q

Type of radiation	Q
X-rays, γ-rays and electrons	1
Thermal neutrons	5
Fast neutrons and protons	20
α-particles	20

Example In one year a worker receives a gamma dose of 0·02 Gy (2 rad), a thermal neutron (N_s) dose of 0·002 Gy (0·2 rad) and a fast neutron dose (N_f) of 0·001 Gy (0·1 rad). What is his total dose equivalent?

Dose equivalent   = absorbed dose × quality factor

γ-dose equivalent = 0·02 × 1   = 0·02 Sv (2.0 rem)

N_s dose equivalent = 0·002 × 5 = 0·01 Sv (1.0 rem)

N_f dose equivalent = 0.001 × 20 = 0·02 Sv (2.0 rem)

Total dose equivalent   = 0·05 Sv (5.0 rem)

3.5 SUBMULTIPLES

In terms of occupational radiation exposure the gray and the sievert are very large units. It is often convenient to have smaller units and this is done by using the prefixes *milli* (one thousandth) abbreviated to m, and *micro* (one millionth) abbreviated to μ. Thus

$$1 \text{ Gy} = 1000 \text{ mGy} = 1\,000\,000 \text{ } \mu\text{Gy}$$
$$1 \text{ Sv} = 1000 \text{ mSv} = 1\,000\,000 \text{ } \mu\text{Sv}$$

In terms of the old units

$$1 \text{ Gy} = 100 \text{ rad}$$
$$1 \text{ mGy} = 100 \text{ mrad}$$
$$1 \text{ } \mu\text{Gy} = 0·1 \text{ mrad}$$

and

$$1 \text{ Sv} = 100 \text{ rem}$$
$$1 \text{ mSv} = 100 \text{ mrem}$$
$$1 \text{ } \mu\text{Sv} = 0·1 \text{ mrem}$$

Example On three successive days a nuclear reactor operator received the following gamma dose equivalent:

     day 1   95 μSv (9·5 mrem)

     day 2   5 μSv (500 μrem)

     day 3   1 mSv (100 mrem)

What was his total dose equivalent in mSv (mrem) over the three days?

Day	Dose equivalent	mSv
1	$95 \ \mu Sv = \dfrac{95}{1000}$ mSv	0·095
2	$5 \ \mu Sv = \dfrac{5}{1000}$ mSv	0·005
3	1 mSv	1·000
Total dose equivalent =		1·100 mSv (110 mrem)

3.6 DOSE RATE

The gray and sievert are units expressing an amount of radiation which may have been received over any period of time. In controlling the radiation hazard it is usually necessary to know the rate at which the radiation is being received. Thus if a man works in an area for 2 hours and receives a dose equivalent of 4 mSv (0·4 rem), then the *dose equivalent rate* is 2 mSv per hour (0·2 rem/h). Similarly *absorbed dose rates* are expressed in Gy/h (rad/h). The relationship between dose, dose rate and time is:

$$dose = dose \ rate \times time$$

Example   If a man is permitted to receive a dose equivalent of 1 mSv (100 mrem) in a week, for how many hours may he work in an area in which the dose equivalent rate is 50 μSv/h (5 mrem/h)?

$$Dose = dose \ rate \times time$$

$$Time = \frac{1 \ mSv}{50 \ \mu Sv/h} = \frac{1000 \ \mu Sv}{50 \ \mu Sv/h} = 20 \ h$$

3.7 FLUX

It is often convenient to express a radiation field as the number of particles or photons crossing an area of 1 square metre in 1 second. This is strictly *fluence rate* but is commonly referred to as *flux* (denoted by Φ). The concept is best illustrated by a practical example.

Consider a point source which emits neutrons at the rate of Q per second (see Fig. 3.3). Now the flux at distance r is the number of neutrons passing through an area of 1 square metre per second. Since the neutrons are being emitted uniformly in all directions, the flux at distance r is the number of neutrons emitted per second divided by the area of the sphere of radius r. This area is $4\pi r^2$ and so the flux Φ is given by:

$$\Phi = \frac{Q}{4\pi r^2} \text{ neutrons per m}^2 \text{ per second (n/(m}^2\text{s))}$$

Note that if r is doubled, r^2 increases fourfold and Φ reduces by four times. This relationship is the *inverse square law* which will be dealt with in greater detail in Chapter 8.

Example Calculate the flux at a distance of 0·5 m from a source which emits 2×10^7 n/s.

$$\Phi = \frac{Q}{4\pi r^2}$$

$$= \frac{2 \times 10^7}{4\pi \times 0\cdot5 \times 0\cdot5}$$

$$= 6\cdot4 \times 10^6 \text{ n/m}^2\text{s}.$$

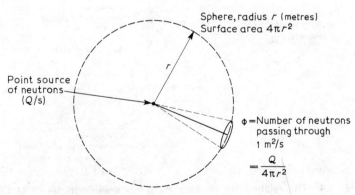

Fig. 3.3 Flux from a point source

Example   Calculate the gamma photon flux at 1 metre from a 0·1 TBq cobalt-60 source. (Cobalt-60 emits 2 gamma rays per disintegration.) From Chapter 2, we know that 0·1 TBq $= 10^{11}$ dis/s, but for ^{60}Co there are two γ-photons per dis. Therefore

$$Q = 2 \times 10^{11} \text{ photons/s}$$

$$\Phi = \frac{Q}{4\pi r^2}$$

$$= \frac{2 \times 10^{11}}{4\pi.1^2}$$

$$= 1 \cdot 6 \times 10^{10} \ \gamma\text{-photons/(m}^2\text{s).}$$

3.8   RELATIONSHIP OF UNITS

The relationship of the units which have now been introduced is illustrated in Fig. 3.4.

The use of the three units, the roentgen, the rad and the rem, was unfortunate and, to the newcomer to health physics, confusing. Whereas exposure, in roentgens, could be used only to describe the effect of X- and γ-rays on air, the rad and rem were applicable to all types of radiation. In SI units, the gray describes an absorbed dose in any medium and the sievert expresses biological effect on the human

Fig. 3.4   The relationship of units (with acknowledgements to Lt Cdr J. Kelly, RN)

body. In health physics, it is clearly the biological effect of radiation which is of interest and so whenever possible dose equivalent in Sv (or rem) should be the unit used.

In everyday health physics, the term *dose* is often loosely used to mean either of the quantities *absorbed dose* or *dose equivalent*. In the following chapters the term *dose* will be taken to mean *dose equivalent*. Absorbed dose and exposure will be given their full titles.

SUMMARY OF CHAPTER

Radiation transfers energy from the source to the absorbing medium.

Ionization: removal of orbital electron – production of ion pairs – in electric field current flows.

Absorbed dose: energy deposition in any medium by any type of ionizing radiation. 1 Gy = 1 J/kg.

Dose equivalent: measure of the biological effect of radiation, unit is the sievert, dose equivalent = absorbed dose × quality factor.

Quality factor: measure of ability of particular type of radiation to cause biological damage, related to density of ionization. $Q = 1$ for β, X and γ, 5 for thermal neutrons, and 20 for fast neutrons and alpha particles.

Dose = dose rate × time.

Flux from point source $= Q/4\pi r^2$.

REVISION QUESTIONS

1. What is ionization and how can it be used as a means of measuring radiation?

2. Discuss the concept of radiation exposure and explain why it is inadequate for the purposes of radiation protection.

3. What is a *gray*? How is it related to the *rad*?

4. What is meant by *dose equivalent*? Explain the relationship between the *sievert* and the *rem*.

5. Explain why the *sievert* is a more suitable unit in health physics than the *gray*.

6. Calculate the neutron flux at a distance of 0·3 m from a neutron source which emits 3×10^7 neutrons per second.

7. Calculate the SI equivalents of the following: 100 mrad, 5 rad, 0·1 mrem, 1 rem.

8. In one week an operator on a nuclear reactor works for 4 hours in an area in which the γ and neutron dose rates respectively are 50 μSv/h and 70 μSv/h, and for a further 18 hours in an area in which the dose rate is 20 μSv/hγ (no neutrons).

Calculate his dose (in sievert) for the week.

4

Biological effects of radiation

4.1 INTRODUCTION

The interaction of ionizing radiation with the human body, arising either from *external* sources outside the body or from *internal* contamination of the body by radioactive substances, leads to biological effects which may later show up as clinical symptoms. The nature and severity of these symptoms and the time at which they appear depend on the amount of radiation absorbed and the rate at which it is received. Radiation injuries can be divided into two classes: *somatic* effects in which the damage appears in the irradiated person himself, and *hereditary* effects which arise only in the offspring of the irradiated person as a result of radiation damage to germ cells in the reproductive organs – the gonads.

4.2 BASIC HUMAN PHYSIOLOGY

Physiology is concerned with the functions of the body as a whole and the component organs and systems. Some basic knowledge of physiology is necessary for an understanding of the ways in which radioactivity can enter and become distributed in the body. Man can be regarded as a machine consisting of various interrelated systems each performing some important function. The systems which are most relevant to an understanding of the behaviour of radioactive substances which enter the body are the circulatory, respiratory and digestive systems (see Fig. 4.1).

4.2.1 The circulatory system

The circulatory system is a closed circuit of tubes around which the blood is pumped by the action of the *heart*. Blood is the transport mechanism of the body and it circulates to almost every region

Fig. 4.1 Schematic illustration of human physiology

carrying food and oxygen to the cells. It also picks up waste products and carbon dioxide and transfers them to the excretory organs. The heart is actually two pumps: the left side pumps the blood through the *arteries* to the tissues. The nourishment is transferred from the tissues to the cells via the tissue fluid. The blood, after passing through the tissues, returns to the right side of the heart via the *veins*. The blood is then pumped to the *lungs* where it becomes oxygenated before returning to the left side of the heart again.

The blood in the arteries contains a lot of oxygen and is bright red in colour while the blood returning from the tissues contains very little oxygen and is dark red. The body contains about 5 litres of blood which circulates about once a minute. There are three types of blood cells, each performing an essential function: the *red cells* (erythrocytes), *white cells* (granulocytes + lymphocytes) and *platelets* (thrombocytes). The function of the red cells is to transport the food and oxygen required by the body, whereas the white cells serve as a means of defence against infection by digesting micro-organisms. Platelets play a vital role in the formation of clots at the site of injuries.

4.2.2 The respiratory system

Respiration (or breathing) is the method by which oxygen is taken into the lungs and carbon dioxide (CO_2) eliminated. The oxygen is absorbed by the blood as it passes through the lungs and carried to the tissues as described above. The tissues produce CO_2 as a gaseous waste product and this is carried back by the blood to the lungs and breathed out. The volume of air breathed per day is approximately 20 cubic metres, of which half is usually considered to be breathed during the eight hours at work.

In the process of respiration, man inhales many 'foreign' substances either in the form of gaseous or particulate materials (i.e. airborne dusts). Gases pass freely into the lungs and enter the bloodstream to a greater or lesser extent depending on their solubility. In the case of particulate matter, only a fraction of the inhaled material is deposited in the lungs, the remainder is either exhaled or deposited in the upper respiratory passages and subsequently swallowed. The behaviour of the material deposited in the lungs depends mainly on its solubility. Highly soluble materials are absorbed rapidly into the bloodstream, perhaps in a matter of hours, whereas insoluble material may persist in the lungs for many months. Clearly, then, the respiratory system represents a route of entry for radioactive substances which can then be transported by the bloodstream to other parts of the body.

4.2.3 The digestive system

The digestive system consists of the oesophagus, the stomach, the duodenum and the small intestine which is connected to the large intestine. Food taken in by the mouth is converted into a form suitable for the production of heat and energy, and the molecules necessary for the growth and repair of tissues. The large molecules in the food are broken down by enzymes in the digestive tract before being absorbed into the bloodstream and passed via the *liver* to the tissues. The unabsorbed food, together with bacteria and cells shed from the intestine wall are passed out as solid waste (faeces). Liquid waste (the waste products of cells dissolved in water) is excreted from the body via the *kidneys* and *bladder* as urine.

Soluble radioactive contamination, when swallowed, may pass through the digestive tract and become absorbed into the blood-stream which carries it to all parts of the body. Insoluble contamination passes through the digestive tract and is excreted in the faeces. During its passage through the body it will irradiate the tract and the large intestine.

4.3   CELL BIOLOGY

All living creatures and organisms consist of tiny structures known as cells. The basic components of a cell are the *nucleus*, a surrounding liquid known as the *cytoplasm* and a *membrane* which forms the cell wall. Figure 4.2 shows a simplified representation of a 'typical' human cell.

The simplest picture of the cell is that the cytoplasm is the 'factory' of the cell while the nucleus contains all the information which the cell needs to carry out its function and reproduce itself. The cytoplasm breaks down food and converts it into energy and small molecules. These small molecules are later converted into complex molecules needed by the cell either for maintenance or duplication.

The nucleus contains the chromosomes which are tiny threadlike structures made up of genes. Human cells normally contain 46 chromosomes. The genes consist of deoxyribonucleic acid (DNA) and

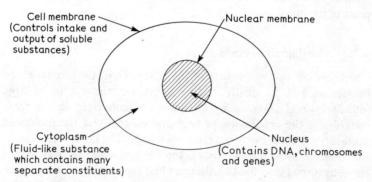

Fig. 4.2   Structure of human cell (schematic)

protein molecules and carry the information which determines the characteristics of the daughter cell.

Cells are able to reproduce to compensate for cells which die. The life of different types of human cell, and hence the rate of reproduction, varies from a few hours to many years. Reproduction of cells occurs in two ways, known as *mitosis* and *meiosis*. The mitotic cells are the ordinary cells in the body and in mitosis the chromosomes duplicate by splitting lengthwise. The original cell then divides into two new cells, each identical to the original cell.

Meiosis is a special kind of division which occurs during the formation of the sexual reproduction cells, namely the sperm in the male and the ovum in the female. It occurs only once in the cell's life-cycle and only in the reproductive cells. In sexual reproduction a sperm and an ovum unite and the chromosomes combine to form a new cell containing genetic material (i.e. genes) from each of the parents. The embryo and subsequently the offspring develops from this single cell – the fertilized ovum.

4.4   THE INTERACTION OF RADIATION WITH CELLS

The basic difference between nuclear radiations and the more commonly encountered radiations such as heat and light is that the former have sufficient energy to cause ionization. In water, of which cells are largely composed, ionization can lead to molecular changes and to the formation of chemical species of a type which are damaging to the chromosome material. The damage takes the form of changes in the construction and the function of the cell. In the human body, these changes may manifest themselves as clinical symptoms such as radiation sickness, cataracts or, in the longer term, cancer.

The processes leading to radiation damage are complex but for many purposes can be considered to take place in four stages:

(a) *The initial physical stage*, lasting only a minute fraction ($\sim 10^{-16}$) of a second in which energy is deposited in the cell and causes ionization. In water the process may be written as:

$$H_2O \xrightarrow{\text{radiation}} H_2O^+ + e^-$$

where H_2O^+ is the positive ion and e^- the negative ion.

(b) *The physico-chemical stage*, lasting about 10^{-6} seconds, in which the ions interact with other water molecules resulting in a number of new products. For example, the positive ion dissociates:

$$H_2O^+ \longrightarrow H^+ + OH$$

the negative ion, that is the electron, attaches to a neutral water molecule which then dissociates:

$$H_2O + e^- \longrightarrow H_2O^-$$
$$H_2O^- \longrightarrow H + OH^-$$

Thus the products of the reactions are H^+, OH^-, H and OH. The two former ions, which are present to quite a large extent in ordinary water, take no part in subsequent reactions. The other two products, H and OH, are called *free radicals*, that is they have an unpaired electron and are chemically highly reactive. Another reaction product is hydrogen peroxide, H_2O_2, which is a strong oxidizing agent and is formed by the reaction:

$$OH + OH \longrightarrow H_2O_2$$

(c) *The chemical stage*, lasting a few seconds, in which the reaction products interact with the important organic molecules of the cell. The free radicals and oxidizing agents may attack the complex molecules which form the chromosomes. They may, for example, attach themselves to a molecule or cause links in long chain molecules to be broken.

(d) *The biological stage*, in which the time scale varies from tens of minutes to tens of years depending on the particular symptoms. The chemical changes discussed above can affect an individual cell in a number of ways. For example they may result in:

  (i) the early death of the cell,
 (ii) the prevention or delay of cell division, or
(iii) a permanent modification which is passed on to daughter cells.

The effects of radiation on the human body are the result of damage to the individual cells. These effects may be conveniently divided into two classes, namely *somatic* and *hereditary*. The somatic effects arise from damage to the ordinary cells of the body and affect only the irradiated person. The hereditary effects, on the other hand, are due to damage to the cells in the reproductive organs, the gonads. The important difference is that, in this case, the damage may be passed on to the person's children and subsequently to later generations.

4.5 THE SOMATIC EFFECTS OF RADIATION

4.5.1 Early radiation effects

The early radiation effects are those which occur in the period from a few hours up to a few weeks after an acute exposure, that is a large dose received over a few hours or less. The effects are due to major depletion of cell populations in a number of body organs due to cell-killing and the prevention or delay of cell division. The main effects are attributable to bone marrow, gastrointestinal or neuromuscular damage depending on the dose received. Acute absorbed doses above about 1 Gy give rise to nausea and vomiting. This is known as radiation sickness and it occurs a few hours after exposure, as a result of damage to cells lining the intestine. Absorbed doses above about 2 Gy can lead to death, probably 10 to 15 days after exposure.

There is no well defined threshold dose below which there is no risk of death due to acute doses, though below about 1·5 Gy the risk of early death would be very low. Similarly, there is no well defined point above which death is certain, but the chances of surviving an acute dose of about 8 Gy would be very low. A reasonable estimate can be made of the dose which would be lethal for 50% of the exposed subjects within 30 days of exposure. This is called the LD_{50}^{30} and is thought to have a value of about 3 Gy for man. For doses of up to about 10 Gy, death is usually due to secondary infections because of depletion of the white blood cells which normally provide protection against infection. The range of doses from 3 to 10 Gy is often called the region of *infection death*. In this range the chances of survival can be increased by special medical treatments, which include isolating the subject in a sterile (i.e. infection free) environment and giving bone marrow transfusion to stimulate white blood cell production.

For doses above about 10 Gy survival time drops abruptly to between 3–5 days. It remains at about this figure until much higher doses are reached. In this region the radiation dose causes severe depletion of the cells lining the intestine. Gross damage occurs in the lining of the intestine, followed by severe bacterial invasion. This is called the region of *gastrointestinal death*.

At much higher doses, survival times become progressively shorter.

There is very little human data in this region but, from animal experiments, the symptoms indicate some damage to the central nervous system; hence the region is called the region of *CNS* (*central nervous system*) *death*. However, it is found that death is not instantaneous even in animals irradiated with doses in excess of 500 Gy.

Another effect which shows up soon after an acute over-exposure to radiation is *erythema*, that is reddening of the skin. Since the skin is located on the surface of the body it is subject to more radiation exposure than most other tissues. This is especially true for beta rays and low energy X-rays. An exposure of about 3 Gy of low energy X-rays will result in erythema and larger exposures may lead to other symptoms such as changes in pigmentation, blistering and ulceration.

The levels of exposure both of workers and of the public arising from the nuclear energy industry, or from industrial and medical applications of radiation, are far below the levels which would induce early effects. Such high doses could only be received in the unlikely event of a major nuclear accident. However, the low doses received in normal operations may cause harmful effects in the long term and these are discussed below.

It will have been noted that in this discussion, early effects have been considered in terms of the absorbed dose, expressed in gray (Gy), rather than as dose equivalent in sievert (Sv). This is really a question of definition; the quality factor, Q, discussed in the previous chapter, and hence the concept of dose equivalent, is intended to apply only to exposures within the normal recommended limits (see Chapter 6) and should not be applied to doses at levels which could lead to early effects.

4.5.2   Late effects

It became apparent in the early part of the twentieth century that groups of people such as radiologists and their patients, who were exposed to relatively high levels of radiation, showed a higher incidence of certain types of cancer than groups not exposed to radiation. More recently, detailed studies of the populations exposed to radiation from atomic bombs, of patients exposed to radiation therapy, and of groups exposed occupationally, particularly uranium miners, have confirmed the ability of radiation to induce cancer.

Cancer is an over-proliferation of cells in a body organ. It is thought that cancer may result from damage to the control system of a single cell, causing it to divide more rapidly than a normal cell. The defect is transmitted to the daughter cells so the population of abnormal cells builds up to the detriment of the normal cells in the organ. The estimation of the increased risk of cancer is complicated by the long and variable latent period, from about 5 to 30 years or more, between exposure and the appearance of the cancer, and by the fact that radiation-induced cancers are not normally distinguishable from those which arise spontaneously. However, at the relatively high levels of exposure of the groups mentioned above, approximate estimates can be made. The extrapolation of these estimates of the risk due to high doses to the much lower levels normally encountered in the nuclear industry and elsewhere introduces major uncertainties. The possibility cannot be ruled out that there is a threshold dose below which there is no risk of radiation-induced cancer. However, this is impossible to demonstrate and it is generally agreed that the only practicable basis for radiological protection is to assume that any dose, no matter how small, carries some risk. This is estimated by extrapolation from the risk at high dose levels on the assumption of a linear relationship between dose and risk. Thus, if the additional cancer risk from a dose equivalent of, say, 100 mSv is one in a thousand, the risk attributable to a dose of 10 mSv would be one in ten thousand. This assumption of a linear relationship forms the basis for the system of dose limits recommended by the International Commission on Radiological Protection (ICRP) – see Chapter 6.

The risks of radiation-induced cancer in the various body organs are under continuous review, as new experimental data become available. Some estimates of the risk of fatal cancers, based on ICRP Publication 26, are shown in Table 4.1. The estimates are expressed as the number of cancer cases expected per million per millisievert. This implies that if one million males each received a dose equivalent of 1 mSv, the number of fatal cancers attributable to the radiation would be about 10, and these would appear over a period of a few decades. In the case of females, the number of radiation-induced cancers would be expected to be about 15 because of the additional risk of breast cancer. The average risk of fatal cancers in a mixed population would be about 13 cases per million per mSv. To put these estimates of the risk of radiation-induced cancer into perspective, it should be noted that in a typical population of 1 million people, there are about 2000 deaths from spontaneous cancers every year.

One of the implications of the assumption of a linear relationship

Table 4.1 Nominal risk of fatal cancer

Effect	Tissue at risk	Cases per 10^6 per mSv Male	Female
Leukaemia	Red bone marrow	2	2
Breast cancer	Female breast	–	5
Lung cancer	Lung	2	2
Thyroid cancer	Thyroid	0·5	0·5
Bone sarcoma	Cells on bone surfaces	0·5	0·5
Other cancers	All other tissues	5	5
Total	Whole body	10	15

between dose equivalent and cancer risk is that the same number of radiation-induced cancers would result from the same dose spread over a different size of population. In the cases discussed above, a dose of 1 mSV to one million people represents a collective dose of 1000 man-Sv. It will be seen that a dose of 0·1 mSv to 10 million people or a dose of 10 mSv to 100 000 people would both correspond to a collective dose of 1000 man-Sv, and so the expected number of fatal cancers due to this dose would be, as before, about 13 cases.

Another possible late effect of radiation is cataract formation in the lens of the eye. In this case it appears that there is a threshold dose, below which cataracts cannot be induced. This is of the order of 15 Sv, and so by setting dose limits so that the total dose to the lens of the eye over the whole working lifetime is below this value, the possibility of cataract formation due to radiation can be avoided.

There is some evidence from animal experiments that exposure to radiation may slightly reduce the life expectation of individuals who do not exhibit any specific radiation-induced symptoms. Observations of human populations exposed at relatively high levels indicate that, if life-shortening occurs at all, it is very slight, almost certainly less than 1 year per sievert.

4.6   THE HEREDITARY EFFECTS OF RADIATION

The hereditary effects of radiation result from damage to the reproductive cells. This damage takes the form of alterations, known as *genetic mutations*, in the hereditary material of the cell.

It has already been mentioned that reproduction occurs when the ovum is fertilized by a sperm. As a result, the offspring receives a complete set of genetic material from each parent. Thus the child receives two complementary sets of genes, one from each of its parents. In general, it is found that one gene is 'dominant' and the other 'recessive'. The dominant gene determines the particular characteristic with which it is associated.

Recessive genes are only recognized when by chance two of the recessive-type genes come together. A considerable number of diseases are associated with recessive genes and will therefore only manifest themselves when both parents have the same recessive genes. Spontaneous mutation accounts for the fact that an appreciable fraction of the world's population suffers from one of the 500 or more defects or diseases attributable to hereditary effects.

Radiation can induce gene mutations which are indistinguishable from naturally-occurring mutations. It should be noted in passing that heat and chemicals can also cause mutations. Mutated genes are generally recessive and so it is generally assumed that all mutations are harmful. This cannot be strictly true since man has attained his present advanced state via a series of mutations. However, this has occurred over an immense time span and the number of harmful mutations which have had to be eliminated from the species over this time is incalculably large.

Since ionizing radiation can cause an increase in the mutation rate, its use will increase the number of genetically abnormal people present in future generations. Clearly, the consequences of excessive genetic damage would be very serious indeed and strict control must be exercised over the radiation exposure of the general population.

The risks of hereditary effects due to exposure of the gonads are very uncertain. ICRP estimate the risk of serious hereditary ill-health within the first two generations following the irradiation of either parent to be about 10 per million per millisievert. Over all generations, the risk would be about twice this value. Clearly, only that exposure which occurs up to the time of conception can affect the genetic characteristics of the offspring, and since the mean age of childbearing is about 30 years, only a proportion of the dose received by a typical population will be genetically harmful. The total genetic risk in all generations, averaged over both sexes and over all ages, is therefore about 8 serious effects per million per mSv.

4.7 DISCUSSION

In this chapter, a distinction has been made between somatic effects, occurring in exposed individuals, and hereditary effects which occur in their descendants. The somatic effects have been discussed under the headings of early and late effects. Recently, another term has been introduced by ICRP to distinguish between effects for which the probability of occurrence depends on the dose received and those for which the severity is related to dose. The former are *stochastic* effects, and the latter *non-stochastic*.

In Publication 26, the ICRP explains that *stochastic* effects are those for which the probability of an effect occurring, rather than its severity, is regarded as a function of dose, without threshold. *Non-stochastic* effects are those for which the severity of the effect varies with the dose, and for which a threshold may therefore occur. Thus cancer induction is a stochastic effect, the probability of a radiation-induced cancer of a particular type depending on the dose received. Hereditary effects are also regarded as being stochastic. The early effects of radiation are non-stochastic since their severity depends on the dose. Similarly, the severity of some late effects, such as cataract formation, depends on the dose received and so these effects are also non-stochastic.

SUMMARY OF CHAPTER

Physiology: the study of the functions of the body as a whole and its
   component organs and systems.
Heart: pumps blood to all parts of the body via the arteries and the
   veins.
Blood: carries food and oxygen to cells and removes waste products
Red blood cells: transport food and oxygen.
White blood cells: defend the body against infection.
Platelets: vital to the formation of clots.
Respiration: method by which oxygen is taken into the lungs and
   carbon dioxide eliminated.
Digestive system: converts food into a form suitable for the produc-

tion of heat and energy, and molecules necessary for the growth and repair of tissues.

Stages in radiation damage process:

- (a) Initial physical stage ($\sim 10^{-16}$ s) consisting of ionization and excitation of atoms and molecules.
- (b) Physico-chemical stage (10^{-8} to 10^{-5} s) consisting of dissociation of ions and formation of free radicals.
- (c) Chemical stage consisting of the interaction of free radicals with other molecules in the body.
- (d) Biological stage in which the chemical reactions show up as effects in individual cells.

Components of cell: nucleus, cytoplasm and outer membrane.

Nucleus: contains chromosomes which are threadlike structures made up of genes.

Genes: carry the information which determines the characteristics of daughter cells.

Mitosis: process by which single cells reproduce.

Meiosis: a stage in the formation of the reproductive cells, the sperm in the male and the ovum in the female.

Effects of radiation on cells: inhibition of mitosis; chromosome aberrations.

Early effects: effects occurring within a few weeks of acute exposure; due to depletion of cell populations.

Late effects: effects occurring at later times, typically some years, after exposure; main effect is cancer induction.

Hereditary effects: appear in descendants of exposed individuals.

Stochastic effects: probability of occurrence depends on dose; mainly cancer and genetic effects.

Non-stochastic effects: severity depends on dose; mainly the early radiation effects plus certain late effects, such as cataract formation.

REVISION QUESTIONS

1. Describe how radioactivity can be deposited in various organs of the body: (a) if it is inhaled, and (b) if it is ingested (swallowed).

2. List the four stages in the radiation damage process.

3. Distinguish between the somatic and hereditary effects of radiation.

4. What are the early radiation effects? Discuss the severity of the effects over the dose range 1 to 10 Gy.

5. What is the major late effect of radiation and upon what assumptions are risk estimates based?

6. What is the difference between stochastic and non-stochastic effects?

5

Natural and man-made radiation

5.1 INTRODUCTION

Throughout history man has been exposed to radiation from the environment. This natural background radiation comes from three main sources – cosmic radiation, radiation from terrestrial sources and radioactivity in the body.

It is impossible to decide whether the natural background radiation has been harmful or beneficial to the development of the human species. It was pointed out in the previous chapter that a very small, but finite, fraction of the natural mutations in cells must be beneficial since it contributes to the evolution of higher forms of life. On the other hand a major proportion of all genetic mutations leads to hereditary defects and genetic death. It is clear that these two effects have achieved some sort of balance and that man has evolved to his present state in spite of background radiation, or perhaps even because of it.

In addition to the natural sources of background radiation many artificial sources of radiation have been introduced in the past 90 years. These artificial sources now add a significant contribution to the total radiation exposure of the population.

5.2 COSMIC RADIATION

Cosmic radiation reaches the Earth from interstellar space and from the Sun. It is composed of a very wide range of penetrating radiations which undergo many types of reactions with the elements they encounter in the atmosphere. The atmosphere acts as a shield and reduces very considerably the amount of cosmic radiation reaching the Earth's surface. This filtering action means that the dose rate at sea-level is less than at high altitudes. For example, the mean dose rate at sea-level, at the equator, is about 0.2 mSv/year (20 mrem/year) while the dose rate at 10 000 feet is about 1 mSv/year (100 mrem/year). The average dose rate in the British Isles from cosmic radiation is about 0·3 mSv/year (30 mrem/year).

One very important radionuclide arises from the interaction of neutrons in cosmic radiation with nitrogen in the upper atmosphere to form carbon-14 as follows:

$$^{14}N(n, p)^{14}C$$

The carbon-14, which has a half-life of 5568 years, diffuses to the lower atmosphere where it may become incorporated in living matter. Similarly, small concentrations of other radionuclides such as ^3H (half-life 12·26 years), ^{36}Cl (half-life $3·08 \times 10^5$ years) and ^{41}Ca (half-life $1·1 \times 10^5$ years) are maintained in the lower atmosphere by cosmic ray reactions. They are much less important than ^{14}C.

5.3   RADIATION FROM TERRESTRIAL SOURCES

The rocks and soil of the Earth's strata contain small quantities of the radioactive elements uranium and thorium with their daughter products. The concentration of these elements varies considerably depending on the type of rock formation. In sandstone and limestone regions the concentration is much lower than in granite. Thus the dose rate from this source depends on the geographical location. In the British Isles, the dose rate to the whole body varies between about 0·2 and 1·0 mSv/year (20 and 100 mrem/year).

Other long-lived isotopes such as ^{48}Ca (half-life $> 7 \times 10^8$ years), ^{50}V (half-life 4×10^{14} years), etc., do not contribute significantly to human dose because they occur in nature in very small concentrations.

5.4   RADIOACTIVITY IN THE BODY

The ingestion and inhalation of naturally occurring radionuclides gives rise to a dose which varies considerably depending on the location, diet and habits of the individual concerned. Potassium-40 and nuclides from the uranium and thorium series contribute most to this dose, with a minor contribution from carbon-14 which originates in the atmosphere. Potassium-40 (^{40}K) is naturally-occurring (half-life $1·27 \times 10^9$ years) and contributes about 0·2 mSv (20 mrem) per year. The dose arising from ingestion of foodstuffs containing trace quantities of members of the uranium and thorium series is about 0.17 mSv (17 mrem) per year.

A significant contribution to the radioactivity in the body comes from the gaseous decay products of the uranium and thorium radioactive series, namely radon and thoron. These gases diffuse from the rocks and soil and are present in easily measurable concentrations in the atmosphere. They are breathed by man along with their decay products and are also taken up by plants and animals with the result that most foodstuffs contain measurable amounts of natural radio-activity. Of ordinary foods, cereals have a high radioactive content while milk produce, fruit and vegetables have a low content. The intake of this type of radioactivity varies greatly with diet and also with location.

Recent studies have shown that there is considerable variability in the doses to members of the UK population from radon and thoron exposure but the average annual dose is estimated to be 0.8 mSv (80 mrem) per year.

5.5   SUMMARY OF DOSES DUE TO NATURAL RADIATION

Table 5.1 gives a list of typical average annual doses due to natural radiation in the British Isles.

Local gamma radiation comes from the ^{238}U and ^{232}Th series, and also from ^{40}K. In certain parts of the world it is much higher than the value given in the above table. For instance, in the monazite sand regions of India and Brazil the annual whole-body doses from local gamma radiation can be as much as 0·12 Sv/year (12 rem/year).

Table 5.1   Typical average annual doses due to natural radiation

Source	Dose ($\mu Sv/year$)
Local gamma radiation	400
Carbon-14	10
Radon, thoron and decay products	800
Potassium-40 in body	200
Cosmic radiation	300
Uranium and thorium nuclides in body	170
	Total: 1880

5.6 MAN-MADE SOURCES OF RADIATION

The early experiences of man-made sources of radiation involved X-rays and various uses of radium. As early as 1896 a letter appeared in *Nature* describing the effects of repeated exposure of the hands to X-rays and during the next 15 years many more cases were reported. These cases arose both from experiments with X-ray sets and also from their use in various treatments. By 1911 Hesse had studied the histories of 94 cases of tumours induced in man by X-rays, of which 50 cases were among radiologists.

These studies illustrated the early types of damage produced by X-rays but gave no indication of the longer-term effects. For some types of damage, such as skin cancer, there is a latent period of between 10–30 years and some radiologists observed malignant skin changes as late as 25 years after discontinuing fluoroscopic examinations. By 1922 it was estimated that more than 100 radiologists had died from occupationally-produced cancer. Similarly, it has been estimated that the death rate from leukaemia among early radiologists was about nine times that among other physicians.

Other studies have shown that the average life-expectancy of the pioneer radiologists was reduced by approximately 2–3 years compared with physicians in general practice. Using results from animal experiments on the relationship between life-shortening and amount of irradiation, it is estimated that the total dose received by the average radiologist from 1935–1958 was a few Gy. The results of these studies have been used by the International Commission on Radiological Protection (ICRP) in deriving the risk factors for irradiation of organs and tissues of the body (see Chapter 6).

Experience was also gained early in this century of the effect of internal dose from various nuclides such as radium (^{226}Ra, half-life 1622 years), mesothorium (^{228}Ra, half-life 5·8 years), radiothorium (^{228}Th, half-life 1·91 years), and their daughter products. Even earlier, it had been recognized that there was a remarkably high incidence of lung cancer among the miners in the Schneeberg cobalt mines of Saxony and the Joachimsthal pitchblende mines in Bohemia. Eventually, this high rate of lung cancer was shown to be due to radiation from the daughter products of uranium, namely ^{226}Ra, ^{222}Rn, ^{218}Po, etc. These mines contain large concentrations of uranium.

During the second and third decades of the twentieth century there were many cases of over-exposure to radium. A considerable number of these arose from the use of radium as a therapeutic agent. It was administered for a large variety of diseases ranging from arthritis to insanity!

The most serious over-exposure to radium occurred in the radium-dial painting industry in the United States. Most of the persons employed were women and they had the habit of 'pointing' their paint brushes with their lips. Many of these women probably ingested tens or even hundreds of microcuries of radium. It is not known how many radium-dial painters actually died from the effects of radiation damage. The study of people exposed to X-rays and radium is continuing to try to establish the degree of risk from acute and chronic exposures to radiation.

5.7 CURRENT SOURCES OF MAN-MADE RADIATION

In addition to the natural background radiation, there are several other sources of human exposure which are peculiar to the last few decades. These sources are: diagnostic radiology, therapeutic radiology, use of isotopes in medicine, radioactive waste, fall-out from nuclear weapon tests, and occupational exposures from nuclear reactors and accelerators.

Diagnostic radiology

It has been estimated that about 90 per cent of the total exposure of the population from medical uses of radiation comes from the diagnostic use of X-rays. The most critical regions of the body are the bone marrow, the gonads and the foetus. The bone marrow is the site of the primitive blood-forming cells and so irradiation of this region could lead to the induction of leukaemia. Irradiation of the gonads is important because of the possibility of genetic damage. Irradiation of pregnant women should be very severely limited because it can lead to physical and mental damage to the child.

Therapeutic radiology

In most countries the average dose to the population from therapeutic radiology is much less than that from diagnostic radiology.

Although quite large exposures may be used in certain treatments, only a small number of people are involved.

Use of radioisotopes

Radioisotopes are used in medicine to give a means of tracing the path and location of specific chemicals in the body. Since radioactive isotopes are chemically identical to stable isotopes of the same element, they will follow the same path and be concentrated to the same degree as the non-active isotopes in the body. By counting methods the location of the active, and hence of the ordinary non-active, isotopes of the element may be determined.

Radioactive waste

The increasing use of radioisotopes and, more particularly, the development of the nuclear power industry results in an ever-growing quantity of radioactive waste. Continued dispersal of low and intermediate levels of radioactive waste to the environment means that members of the general population will receive an increasing exposure from this source. At present the contribution to the total exposure of the population from waste disposal is very low, about 2 μSv/year (0.2 mrem/y), but it seems certain to increase as more power stations are built. For this reason very strict control is exercised over the dispersal of radioactive waste to the environment. The problems of radioactive waste disposal are discussed in Chapter 11.

Fall-out from weapons testing

The nuclides of concern in radioactive fall-out from nuclear weapons testing are similar to those arising from the operation of nuclear power stations. The two most important radionuclides are strontium-90 (^{90}Sr, half-life 28.8 years) and caesium-137 (^{137}Cs, half-life 30.0 years). Strontium-90 concentrates in the skeleton and caesium-137 is distributed uniformly throughout the body.

Some of the radionuclides created during a nuclear test are injected into the troposphere (40 000–60 000 feet) and are carried around the Earth several times. They gradually return to the Earth over a period of a few years and consequently give appreciable doses to the world's population. The dose from nuclides injected into the troposphere reaches a peak shortly after each weapon test.

Occupational exposure

The dose from all occupational exposure, both research and production, is very small when averaged over the whole population. The estimated dose in the United Kingdom is about 9 μSv/year (0·9 mrem/year), of which atomic energy workers contribute about 40 per cent. The remainder is due to uses of radiation in industry and medicine.

5.8   SUMMARY OF CURRENT SOURCES OF RADIATION

Table 5.2 lists the average annual doses received by members of the public in the UK from the current sources of man-made radiation.

SUMMARY OF CHAPTER

Sources of background radiation: cosmic radiation, radiation from uranium and thorium with their daughter products, radioactivity in the body.

Man-made sources of radiation: diagnostic radiology, therapeutic radiology, use of radioisotopes in medicine, radioactive waste,

Table 5.2   Average annual doses due to man-made radiation in the UK

Source	Dose ($\mu Sv/year$)
Diagnostic radiology	220
Therapeutic radiology	30
Use of isotopes in medicine	2
Radioactive waste	2
Fall-out from nuclear weapons	10
Occupationally exposed persons	9
Other sources	12
Total: approx.	285

fallout from weapons testing, occupational exposure from nuclear reactors and industrial applications.

REVISION QUESTIONS

1. List the main sources of natural background radiation and discuss how these sources vary with:

   (*a*) altitude,
   (*b*) geographical location.

2. Calculate the average dose received by members of the general population (from background radiation) in the United Kingdom over the first 30 years of their life.

3. Explain the difference between diagnostic and therapeutic radiology and comment on the contribution of each to the average dose received by members of the public.

4. Discuss the probable way in which the dose to members of the public from the disposal of radioactive waste will vary with time.

The system of
dose limitation

6.1 THE ROLE OF THE ICRP

The International Commission on Radiological Protection (ICRP) was established by the second International Congress of Radiology (ICR) in 1928. Since its inception, the ICRP has been the one internationally recognized body responsible for recommending values of dose limits for exposure to ionizing radiation. Most of the national legislation relating to exposure to radiation is based on the recommendations of the ICRP.

The early recommendations of ICRP were concerned with protection against X-rays and radium. Some of the earliest recommendations dealt with the length of time that a worker should be engaged on radiation work; these were:

(a) Not more than 7 h/day.
(b) Not more than 5 days/week.
(c) Not less than 1 month's holiday per year.
(d) Off days to be spent as much as possible out of doors.

The maximum permissible doses were defined very loosely.

In 1950, the ICRP extended its scope to deal with the many new problems resulting from the introduction of nuclear reactors. The dose limits recommended by ICRP have changed considerably over the past five decades, as shown in Table 6.1.

All of these changes have led to a decrease in the occupational exposure values. This continuous decrease has not been due to any positive evidence of damage to persons working within the earlier permissible dose levels, but rather to an increasing awareness of the uncertainties in much of the available experimental data and hence of the need for caution in setting limits.

In early ICRP recommendations the expression *tolerance dose* was used to describe the acceptable level of exposure to radiation. The

Table 6.1   History of dose limits for workers

Dose rate	Date recommended	Comments
0·1 of an erythema dose per year	1925	Proposed by A. Mutscheller and R.M. Sievert. This corresponds to an exposure of ∼30 R/year from 100 kV X-rays or ∼70 R/year from 200 kV X-rays
0·2 R/day or 1 R per working week	1934	Recommended by ICRP
150 mSv/year (15 rem/year), or approximately 3 mSv/week (0·3 rem/week)	1950	Recommended by ICRP
50 mSv/year (5 rem/year), or approximately 1 mSv/week (0·1 rem/week)	1956	Recommended by ICRP
All exposures to be kept as low as reasonably achievable; dose equivalent limit 50 mSv (5 rem) in a year	1977	Recommended by ICRP

term tolerance dose had the unfortunate connotation that it seemed to imply a threshold dose below which no radiation damage would occur. On the basis of presently available evidence there is no proof of the existence of a threshold dose for certain types of somatic damage. Consequently, the term tolerance dose was replaced by the term *maximum permissible dose* (MPD). In 1958, the ICRP gave the following definition of MPD:

The permissible dose for an individual is that dose, accumulated over a long period of time or resulting from a single exposure, which in the light of present knowledge carries a negligible probability of severe somatic or genetic injuries; furthermore, it is such a dose that any effects that ensue more frequently are limited to those of a minor nature that would not be considered unacceptable by the exposed individual and by competent medical authorities. Any severe somatic injuries (e.g. leukaemia) that might result from exposure of individuals to the permissible dose would be limited to an exceedingly small fraction of the exposed group; effects such as shortening of life span, which might be expected to occur more frequently, would be very slight and would likely be hidden by normal biological variations. The permissible doses can therefore be expected to produce effects that could be detected only by statistical methods applied to large groups.

In its most recent recommendations (ICRP Publication 26, 1977) the Commission no longer uses the phrase *maximum permissible dose*. Instead it recommends a more comprehensive system of dose limitation, the main features of which may be summarized as follows:

(a) no practice involving radiation exposure shall be adopted unless its introduction produces a net benefit;

(b) in addition, all exposures should be kept as low as is reasonably achievable, economic and social factors being taken into account; and

(c) the dose equivalent to individuals shall not exceed the limits recommended for the appropriate circumstances by the Commission.

Clearly the emphasis is now on ensuring that all exposures are kept as low as is reasonably achievable with the dose equivalent limit as a final 'backstop'. The detailed recommendations of ICRP Publication 26 and subsequent statements by the Commission are discussed in the next section.

6.2 RECOMMENDATIONS OF ICRP (PUBLICATION 26, 1977)

In its Publication 26, ICRP differentiates between two different types of effects which may be induced by ionizing radiation.

6.2.1 Stochastic effects

These are those effects for which the probability of an effect occurring, rather than its severity, is regarded as a function of dose, without threshold. The most important somatic stochastic effect is the induction of cancers, for which the risk must be regarded as increasing progressively with increasing dose received, without threshold. Similarly, at the dose levels involved in radiation protection, genetic effects are regarded as being stochastic.

6.2.2   Non-stochastic effects

These are effects for which the severity of the effect varies with the dose, and for which a threshold may therefore exist. Examples of non-stochastic injuries are cataract of the lens of the eye, damage to blood vessels and impairment of fertility. The severity of these effects varies with the size of the radiation dose received but they are not detectable at all unless a quite high threshold dose is exceeded.

The aim of radiation protection, as stated by ICRP, should be to prevent detrimental non-stochastic effects and to limit the probability of stochastic effects to levels deemed to be acceptable. This aim is achieved by:

(a) setting dose equivalent limits at levels which are sufficiently low to ensure that no threshold dose is reached, even following exposure for the whole of an individual's lifetime – *prevention of non-stochastic effects*; and

(b) keeping all justifiable exposures as low as is reasonably achievable, economic and social factors being taken into account, subject always to the boundary condition that the appropriate dose equivalent limits shall not be exceeded – *limitation of stochastic effects*.

6.3   RECOMMENDED DOSE EQUIVALENT LIMITS FOR WORKERS

To prevent non-stochastic effects a dose equivalent limit of 0·5 Sv (50 rem) in a year is recommended for all tissues except the lens of the eye, for which a lower limit of 0·15 Sv (15 rem) in a year is recommended.

To limit stochastic effects the annual dose equivalent limit for uniform irradiation of the whole body is set at 50 mSv (5 rem). For non-uniform irradiation of the body, weighting factors have been assigned to the various individual organs, relative to the whole body as 1·0, reflecting the harm attributable to irradiation of each organ. The sum of the weighted organ doses is known as the *effective dose equivalent*, H. Thus

$$H = \sum_T w_T H_T$$

where w_T is the weighting factor for tissue T and H_T is the dose equivalent in tissue T. The annual limit on effective dose equivalent is 50 mSv (5 rem) and so in any one year

$$\sum_T w_T H_T \leq 50 \text{ mSv (5 rem)}$$

The weighting factors are given in Table 6.2.

Table 6.2 Tissue weighting factors

Tissue	w_T
Gonads	0·25
Breast	0·15
Red bone marrow	0·12
Lung	0·12
Thyroid	0·03
Bone surfaces	0·03
Remainder	0·30

Example 1 Calculate the allowable dose equivalent to the thyroid of a worker for a year in which he is exposed to non-uniform irradiation involving the whole body and the lung, as well as the thyroid. During the year he receives dose equivalents of 25 mSv (2·5 rem) to the whole body and 150 mSv (15 rem) to the lung.

Using the weighting factor formula

$$\sum_T w_T H_T \leq 50 \text{ mSv (5 rem)}$$

w (whole body) $\times H$ (whole body) $+ w$ (lung) $\times H$ (lung) $+$ w(thyroid) $\times H$ (thyroid) ≤ 50 mSv (5 rem)

Thus

$$1·0 \times 25 \text{ mSv} + 0·12 \times 150 \text{ mSv} + 0·03 \times H \text{ (thyroid)}$$
$$= 50 \text{ mSv, in the limit}$$

i.e.

$$25 \text{ mSv} + 18 \text{ mSv} + 0.03 \, H \text{ (thyroid)} = 50 \text{ mSv}$$

$$H \text{ (thyroid)} = \frac{50 - 43}{0.03} = 233 \text{ mSv}$$

Thus the worker is permitted to receive up to 233 mSv (23.3 rem) dose equivalent to the thyroid during the year in question.

Example 2   Using the weighting factors in Table 6.2, calculate the implied limits for each of the following organs, assuming that each organ is irradiated completely in isolation: the gonads, the thyroid, the bone surfaces.

For gonads, $w_T = 0.25$ and so

$$\text{implied annual limit} = \frac{50}{0.25} = 200 \text{ mSv (20 rem)}$$

For thyroid, $w_T = 0.03$, thus

$$\text{implied annual limit} = \frac{50}{0.03} = 1670 \text{ mSv (167 rem)}$$

For bone surfaces, $w_T = 0.03$ and so

$$\text{implied annual limit} = \frac{50}{0.03} = 1670 \text{ mSv (167 rem)}$$

Note: The above limits are based on stochastic effects. For all organs and tissues there is an additional limitation, based on non-stochastic effects, which prohibits any organ or tissue receiving an annual dose equivalent greater than 500 mSv (50 rem) (or 150 mSv (15 rem) for lens of eye). Thus, while the annual dose equivalent limit for gonads is the stochastic limit of 200 mSv (20 rem), the annual dose equivalent limit for both the thyroid and the bone surfaces is the non-stochastic limit 500 mSv (50 rem).

6.4   NOTES ON THE DOSE EQUIVALENT LIMITS FOR WORKERS

The following points should be stressed in applying this system of dose limitation:

(a) All unnecessary exposures should be avoided.

(b) The Commission previously recommended a separate dose limit for any one quarter, in addition to the annual dose limit. In Publication 26, however, the Commission recommends only annual dose equivalent limits.

(c) The Commission lays considerable emphasis on the fact that only a few workers would be expected to receive annual dose equivalents close to the recommended limit. Experience shows that in many industries the distribution of doses has often been such that the average worker has received an annual whole body dose equivalent of about 5 mSv (500 mrem). Using the risk factors quoted in Publication 26 this implies that the average risk of death in such radiation occupations is comparable with the average risk in other industries which are normally considered 'safe'.

(d) ICRP defines two conditions of work:

Working condition A, where the annual exposures might exceed 3/10 of the dose equivalent limits. Category A workers should be subject to individual monitoring for external irradiation and for internal contamination as appropriate. They should also be given a pre-operational medical examination.

Working condition B, where it is most unlikely that the annual exposures will exceed 3/10 of the dose equivalent limits. Category B workers are not required to have a pre-operational medical examination. In many circumstances they may be subject to personnel monitoring to confirm that control is satisfactory.

6.5 RECOMMENDED DOSE EQUIVALENT LIMITS FOR INDIVIDUAL MEMBERS OF THE PUBLIC

In March 1985, the ICRP issued a statement in which they emphasized that the principal limit for members of the public is an annual effective dose equivalent of 1 mSv (100 mrem). However, they recommended that it is permissible to use a subsidiary annual dose limit of 5 mSv (500 mrem) for some years, provided that the lifetime average annual effective dose equivalent does not exceed 1 mSv (100 mrem). In these circumstances, it is unnecessary to specify dose equivalent

limits for non-stochastic effects, although in some situations the lens
of the eye and the skin may require special consideration. To cope
with such circumstances, the ICRP recommends an annual dose
equivalent limit of 50 mSv (5 rem) for each.

6.6 PLANNED SPECIAL EXPOSURES

The ICRP recognizes that very occasionally situations may occur,
during normal operations, when it may be necessary to permit a few
workers to receive dose equivalents in excess of the recommended
limits. In such circumstances, where no alternative procedures are
available or practicable, external exposures or intakes of radioactive
material may be permitted provided the dose equivalent commitment
does not exceed twice the relevant annual limit (e.g. 100 mSv (10 rem)
to the whole body) in any single event, or five times this limit (e.g.
250 mSv (25 rem) to the whole body) in a lifetime. Planned special
exposures should not be permitted to a worker who has previously
received abnormal exposures in excess of five times the relevant
annual limit, or to women of reproductive age.

6.7 ABNORMAL EXPOSURES IN EMERGENCIES OR
ACCIDENTS

It is obvious that no generally applicable dose limit can be specified
for accidental exposures. The ICRP states that it is not possible for
them to recommend *intervention levels* that would be appropriate in
all circumstances.

To limit exposure of workers and the general public following an
accidental release of radioactive material it is necessary to have a
detailed and well rehearsed emergency plan. The emergency plan has
three objectives:

(*a*) to restrict exposures as far as is reasonably achievable and, in
       particular, to attempt to avoid exposures above the dose equi-
       valent limits;
(*b*) to bring the situation back under control;
(*c*) to obtain information for assessing the causes and conse-
       quences of the incident.

In an emergency, volunteers may receive large doses for the purpose of saving life or preventing serious injuries, or to prevent a substantial increase in the scale of the incident. It is difficult to specify limits to cover such an event since each situation will be unique. However, it may be that some volunteers will have to be exposed in excess of the limits applicable to a planned special exposure, especially where the saving of life is involved. When rescue operations have to be undertaken, it may not be possible to work to specified dose limits. Each situation must be assessed by the competent person or persons, and a decision reached on the basis of this assessment.

It is probable that whole-body doses of up to 1 Gy (100 rad) might be acceptable in the saving of life. If the operation would require doses much in excess of this level, then the risks and possible result of the operation would have to be judged very carefully. One important consideration would be the accuracy of the information regarding the likely dose rates in the accident area; a second would be the condition of the casualties and their likelihood of survival. For instance, if the estimates of dose rate were low by a factor of two, a rescuer might unwittingly receive a sufficiently large dose to produce severe early somatic effects.

Regarding exposure of members of the public following an accident, the ICRP believes that it is not possible for them to fix generally applicable intervention levels above which intervention by the appropriate authorities would always be required. All countermeasures that can be applied to reduce the exposure of members of the public following an accidental release of radioactive materials carry some detriment to the people concerned. Thus the decision to introduce countermeasures (i.e. make an intervention) must be based on a balance of the detriment which they carry and the reduction in the exposure that they are likely to achieve.

However, the ICRP judge that it might be possible to set levels below which intervention would not generally be considered to be justified. Such intervention levels and derived intervention levels should be included in the emergency plan, which should also give guidance on the different countermeasures available. It is emphasized that the emergency plan should be sufficiently flexible to allow its adaption to the real accident situation. In particular, the intervention levels should not be applied in an automatic fashion but should be reassessed in the light of the available information at the time of intervention.

SUMMARY OF CHAPTER

International Commission on Radiological Protection (ICRP): internationally recognized body responsible for recommending a system of dose limitation.

Stochastic effects: those for which the probability of an effect occurring, rather than its severity, is regarded as a function of dose, without threshold (e.g. cancer induction, genetic effects).

Non-stochastic effects: those for which the severity of the effect varies with the dose, and for which a threshold may therefore apply (e.g. cataracts of the lens of the eye, damage to blood vessels, impairment of fertility).

Aim of ICRPs recommendations: to prevent detrimental non-stochastic effects and to limit the probability of stochastic effects to acceptable levels.

Non-stochastic dose equivalent limit: 0·5 Sv (50 rem) in a year for all tissues except lens of the eye, for which limit is 0·15 Sv (15 rem) in a year.

Stochastic dose equivalent limit: 50 mSv (5 rem) in a year to the whole body; for non-uniform irradiation, apply formula

$$\sum_T w_T H_T \le 50 \text{ mSv (5 rem)}$$

with the appropriate weighting factors (w_T) for individual organs.

Working condition A: where annual exposures might exceed 3/10 of the dose equivalent limits, individual monitoring and preoperational medical examination required.

Working condition B: where annual exposures are unlikely to exceed 3/10 of the dose equivalent limits, individual monitoring and preoperational medical examination not necessary.

Dose equivalent limits for individual members of the public: lifetime annual average effective dose equivalent limit is 1 mSv (0·1 rem). Subsidiary annual dose limit of 5 mSv (0·5 rem) for some years, subject to lifetime average not being exceeded.

Planned special exposures: up to twice the relevant annual dose equivalent limit in any single event, or five times the limit in a lifetime, under very special circumstances.

Emergency exposures:

(a) attempt to avoid exposures above the dose equivalent limits;

(b) saving of valuable equipment, up to the limits for planned special exposures;

(c) saving of life, may be justifiable to exceed the limits for planned special exposures, probably up to about 1 Gy (100 rad);

(d) intervention levels to reduce exposures of public.

REVISION QUESTIONS

1. Explain the main features of the system of dose limitation recommended by ICRP in its Publication 26.

2. Explain what is meant by the terms *stochastic* and *non-stochastic* effects and give two examples of each type of effect.

3. Give the annual non-stochastic limit (for workers) for each of the following: the lung, the thyroid, the lens of the eye.

4. Explain how the dose equivalents to various organs of the body from non-uniform irradiation are related to the whole-body limit for uniform irradiation.

5. A worker is required to work in an area where he or she is subjected to non-uniform irradiation, involving exposure of the whole body, the red bone marrow and the lung. During one year the following dose equivalents are received:

whole body	10 mSv (1 rem)
lung	100 mSv (10 rem)
red bone marrow	150 mSv (15 rem)

Calculate the effective dose equivalent commitment.

6. (a) Assuming, in turn, that each of the following organs of a worker is irradiated for the entire year in isolation, calculate the annual dose equivalent limit implied for each organ by the weighting factor formula: the breast, the red bone marrow, the lung and the thyroid.

    (b) Write down the non-stochastic dose equivalent limits for each of the above organs.

    (c) State, for each organ, the governing annual dose equivalent limit.

7. State the main differences between working conditions A and B.

8. Explain the difference between a *planned special exposure* and an exposure in an accident or an emergency. Why might it sometimes be permissible, following an accident, for workers to be exposed in excess of the planned special exposure limits?

Radiation
detection and measurement

7.1 GENERAL PRINCIPLES

The fact that the human body is unable to sense ionizing radiation is probably responsible for much of the general apprehension about this type of hazard. Reliance must be placed on detection devices which are based on the physical or chemical effects of radiation. These effects include:

(*a*) ionization in gases,
(*b*) ionization and excitation in certain solids,
(*c*) changes in chemical systems, and
(*d*) activation by neutrons.

The majority of health physics monitoring instruments use detectors based on ionization of a gas. Certain classes of crystalline solids exhibit increases in electrical conductivity and effects attributable to excitation, including scintillation, thermoluminescence and the photographic effect. Detection systems are available in which chemical changes are measured but these are rather insensitive. A method which may be applied to neutron detection depends on the activation caused by neutron reactions.

In this chapter, the basic principles of those systems commonly used in practical health physics are described. Their applications to particular types of measurement are covered in Chapters 8, 9 and 15.

7.2 IONIZATION OF A GAS

7.2.1 Ionization chamber

It will be recalled from Chapter 3 that the absorption of radiation in a gas results in the production of *ion pairs* consisting of a *negative ion* (the electron) and a *positive ion*. A moderate voltage applied between two plates (electrodes) in close proximity causes the negative ions to be attracted to the positive electrode (anode) and the positive ions to the negative electrode (cathode). This flow of ions constitutes an electric current which is a measure of the intensity of radiation in the gas volume. The current is extremely low (about 10^{-12} A) and a sensitive electronic circuit known as a d.c. (direct current) amplifier is used to measure it. This system is known as an ionization chamber and the current measured is a mean value due to the interaction of many charged particles or photons (Fig. 7.1).

The design of the chamber and the filling gas used depends on the particular application. In health physics instruments the chamber is usually air filled and is constructed of low atomic number materials. If the instrument is required to respond to beta radiation the chamber must have thin walls or a thin window.

7.2.2 Proportional counter

If, in an ion chamber system, the applied voltage is increased beyond a certain point, an effect known as *gas amplification* occurs. This is because the electrons produced by ionization are accelerated by the applied voltage to a sufficiently high energy to cause further ionization themselves before reaching the anode, and a cascade of ionization results (Fig. 7.2). Thus a single ionizing particle or photon can produce a pulse of current which is large enough to be detected. Over a certain range of voltage the size of the pulse is proportional to the

Fig. 7.1 Ion chamber system

Fig. 7.2 Gas amplification

amount of energy deposited by the original particle or photon and so the system is known as a proportional counter. The term *counter* means that the output is a series of pulses, which may be counted by some means, rather than an average current as obtained with an ionization chamber.

7.2.3 Geiger-Müller counter

If the voltage in the ionization system is increased still further the gas amplification is so great that a single ionizing particle produces an avalanche of ionization resulting in a very large pulse of current. The size of the pulse is the same, regardless of the quantity of energy initially deposited, and is governed more by the external circuit than the counter itself. The G-M tube is very widely used in monitoring equipment because it is very rugged and can operate relatively simple output circuits. Again, this is a counting device but it is possible in some cases to use a G-M counter in a circuit which measures the average current flowing through the tube.

In practice both proportional and G-M counters are usually con-structed in the form of a cylinder which forms the cathode with a central thin wire which is the anode. The whole is enclosed in a glass or metal tube which is filled with a special gas mixture.

7.3   SOLID STATE DETECTORS

7.3.1   Mechanism

The term *solid state detectors* refers to certain classes of crystalline substances which exhibit measurable effects when exposed to ionizing radiation. In such substances, electrons exist in definite *energy bands* separated by *forbidden bands*. The highest energy band in which electrons normally exist is the *valence band*. The transfer of energy from a photon or charged particle to a valence electron may raise it from the valence band through the forbidden band into the *exciton* or *conduction band*. The vacancy left by the electron is known as a *hole* and it is analogous to a positive ion in a gas system.

The raising of an electron to the conduction band is known as ionization and the electron–hole pair can be compared with ion pairs in a gas. The electron and hole are independently mobile and in the presence of an electrical potential will be oppositely attracted, so contributing to electrical conduction in the material. If an electron is raised to the exciton band, the process is excitation. In this case the electron is still bound to the hole by electrical forces and so they cannot contribute to conduction. The third process which can occur is *electron trapping*. Traps are imperfections or impurity atoms in the crystal structure which cause electrons to be caught in the forbidden band. The three processes are illustrated in Fig. 7.3.

Fig. 7.3   Ionization, excitation and trapping

The existence of the three states may be virtually permanent or they may last a very short time depending on the material and, to a great extent, the temperature. In returning to the valence band, the difference in energy is emitted as *fluorescent* radiation, usually a photon of visible light. In the case of trapped electrons, energy must first be provided to enable the electron to escape from the trap back into the exciton band and thence down to the valence band. The energy to release the electrons is usually provided by raising the temperature of the substance; the light given off as a result is known as *thermoluminescence*.

The practical application of the three processes of conductivity, fluorescence and thermoluminescence will be considered in more detail below. It should be mentioned that the photographic effect is a solid state process but it will be treated separately.

7.3.2 Conductivity detectors

Since changes in conductivity are due to ionization, solid state conductivity detectors are similar in some ways to gas ionization systems. A cadmium sulphide (CdS) detector, for example, is analogous to an ion chamber. It is operated in the mean current mode and is suitable in some applications for the measurement of γ dose rate. The main advantage is that it can be much smaller than an ion chamber and yet have a greater sensitivity.

As with gas systems, some solid state detectors, notably germanium and silicon, operate in the pulse mode. Germanium has the disadvantage that it must be operated at very low temperatures. The pulse size in both cases is proportional to the energy deposition within the detector. The main application is in gamma spectrometry, in which by analysing the size of pulses from the detector it is possible to measure the energy of γ-rays.

7.3.3 Scintillation detectors

Scintillation detectors are based on detection of the fluorescent radiation (usually light) emitted when an electron returns from an excited state to the valence band. The material selected is one in which this occurs very quickly (within about 1 μs). The absorption of a 1 MeV γ-photon in a scintillation detector results in about 10 000 excitations and the same number of photons of light. These *scintillations* are detected by means of a photomultiplier tube which converts the light into electrical pulses which are then amplified. The size of pulse is proportional to the energy deposited in the crystal by the charged particle or photon. In γ-ray work the most common scintillator is sodium iodide (NaI) usually in crystals of about 50×50 mm. These are widely used in γ-spectrometry but are being supplanted by germanium detectors which offer better energy resolution. Zinc sulphide crystals in very thin layers are used for α-detection.

7.3.4 Thermoluminescence detectors

Thermoluminescence detectors (TLD) utilize the electron trapping process. The material is selected so that electrons trapped as a result of exposure to ionizing radiation are stable at normal temperatures. If, after irradiation, the material is heated to a suitable temperature, usually about 200 °C, the trapped electrons are released and return to the valence band with the emission of a light photon. Thus if the device is heated in the dark but under a photomultiplier tube, the light output can be measured and this is proportional to the radiation dose which the detector has received. The most common material used is lithium fluoride (LiF) but various other materials including calcium fluoride and lithium borate have special applications.

It should be noted that whilst the methods previously described are more suitable for measurement of radiation intensity (i.e. dose *rate*), the TLD system measures the total dose accumulated over the period of exposure.

7.4 PHOTOGRAPHIC EFFECT

Ionizing radiation affects photographic film in the same way as does visible light. A photographic film consists of an emulsion of crystals (grains) of silver bromide on a transparent plastic base. The absorption of energy in a silver bromide grain, whether from light or ionizing radiation, results in the formation of a small cluster (often only a few atoms) of metallic silver. This cluster is known as a *latent image*. When the film is *developed* this tiny amount of silver assists the conversion of all the silver in a grain from its compound form, silver bromide, into metallic silver which deposits on the plastic base material. This is an amplification process giving a gain of about 10^9 which accounts for the high sensitivity of photographic emulsions. After development the film is *fixed* or made stable by washing in a sodium thiosulphite (hypo) bath which removes any unconverted silver bromide. If good results are to be obtained it is important to control strictly developer strength, temperature and processing time.

Photographic films used for radiation monitoring are usually 30×40 mm and are, of course, sealed in a light-tight packet. After processing, the film is read by passing a beam of light through it and measuring the optical density. This observed density is converted to radiation dose by means of a calibration curve obtained by exposing a number of films to known doses and plotting a dose–density curve (see Fig. 7.4).

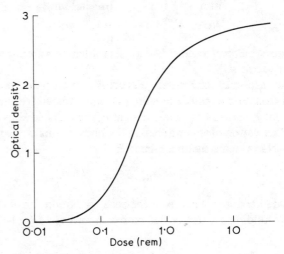

Fig. 7.4 Dose-density curve.

The sensitivity of the film depends on the grain size of the emulsion. The most sensitive types give a range of dose measurement of about 50 μSv to 50 mSv (5 mrem to 5 rem).

The main advantage of photographic film is that, with the aid of special film holders incorporating filters, it enables information on the type and energy of radiation to be deduced. In addition, the developed film can be stored and rescrutinized later. The most serious disadvantage is that a rapid reading cannot be obtained.

7.5   ACTIVATION EFFECT

The bombardment of most elements by neutrons produces radio-active nuclides and measurement of the degree of activation permits estimation of the incident neutron flux. In most cases the method is not very sensitive and the main application is in the assessment of large accidental doses.

A method suitable for fast neutron measurement is the use of sulphur discs which undergo the reaction.

$$^{32}S(n, p)^{32}P \qquad (P-phosphorus)$$

Other useful reactions include

$$^{115}In(n, \gamma)^{116}In \qquad (In-indium)$$
$$^{197}Au(n, \gamma)^{198}Au \qquad (Au-gold)$$

The nuclides ^{32}P, ^{116}In and ^{198}Au are β-emitters and are counted in a suitable system.

Another aspect of the activation effect is that a person receiving a large neutron dose would be rendered slightly radioactive and a dose estimate may be made by measurement of the induced activity. For example activation of sodium (Na) in the body results in the production of ^{24}Na, which is again a β-emitter.

$$^{23}Na(n, \gamma)^{24}Na$$

With moderate doses of neutrons the decay radiation can be detected by simply holding a sensitive detector such as a G-M probe against the body.

7.6   ELECTRICAL CIRCUITS

7.6.1   Types of circuit

In Sections 7.2 and 7.3 various detectors were described which give as output an electrical signal. These electrical signals are of two types: direct current and pulse. Measurement of direct current is by d.c. amplifiers whilst for pulsed output, counting circuits or ratemeters are used. The important practical features of these circuits are described below.

7.6.2 d.c. amplifier

A d.c. amplifier is a means of amplifying a very low current to a high enough value to operate a conventional ammeter. In ion chamber systems the *gain* required is quite high, the current input may be about 10^{-12} A whilst about 10^{-3} A is necessary to drive a meter. The required gain in this case is 10^9. Amplifiers with such a high gain have a tendency to instability due to temperature fluctuations, etc. This can be reduced to a marked extent by using *negative feedback* (see Fig. 7.5). Briefly this means that if an increased signal is fed into the input, it results in an opposite signal being fed from the output back to the input. Another important point is that when currents as low as 10^{-12} A are being measured, great care must be taken to avoid stray currents affecting the input. A very high standard of insulation and cleanliness is required between the detector output and the input to the amplifier.

The main application of d.c. amplifiers is in ion chamber systems, both portable and fixed, but they are also used for solid state detectors of the CdS type.

Fig. 7.5   d.c. amplifier – negative feedback

7.6.3 Pulse counting systems

A basic counting system consists of a *pulse amplifier*, a *discriminator* and a *scaler*. In addition the system may contain an *EHT* (extra high tension) *unit* to provide a high voltage supply for the detector.

The function of the *pulse amplifier* is to accept pulses from the detector and amplify them to a size compatible with subsequent circuits. The size of pulses depends on the type of detector but is typically a few μV (microvolts) for solid state detectors, a few mV for proportional and scintillation counters up to a few volts for G-M counters. Since most counting circuits operate on pulses of a few volts it will be seen that the amplifier requires a gain of about 10^6 for solid state detectors, 10^3 for proportional and scintillation counters and 1 for G-M counters.

In all electronic apparatus there is present, to some extent, electronic 'noise' in the form of small electrical pulses. These noise pulses are amplified with the signal pulses from the detector and, unless precautions are taken, will be counted by the system. The function of the *discriminator* is to reject all pulses below a certain level, which is set by applying a *discriminator bias* voltage. The equipment will then record only those pulses whose amplitude exceeds the bias level. Figure 7.6 shows a train of pulses and small noise pulses being fed from an amplifier to a discriminator. If the bias level is reduced below A, noise pulses will be counted and if it is raised above C, detector pulses will not be counted. The correct level is that indicated by B. A discriminator bias characteristic can be plotted by measuring the count rate from a detector with a range of discriminator bias settings. This is illustrated in Fig. 7.7 in which the lines A, B and C correspond to those in the previous figure. Thus if the bias voltage is below A a very high count rate is recorded, if it exceeds C, no counts are recorded. The correct setting is at B at which only genuine pulses from the detector will be recorded. In addition, since the curve is almost level at this point small variations in the bias setting will not seriously affect the count rate.

The *scaler* accepts pulses from the discriminator and gives a visual display of the number of pulses (counts) received over the counting period. The scaler usually incorporates a timing device so that once started it will count for a pre-set time ranging from a few seconds up to a few hours.

Fig. 7.6 Function of a discriminator

Fig. 7.7 Discriminator bias characteristic

The counting rate from detectors is dependent on the voltage applied. In order to set up the equipment, a small source is placed near the detector and a series of counts made for different detector voltages. The graph obtained by plotting these results is called a *plateau* because the count rate is relatively independent of the applied voltage over a certain range (see Fig. 7.8). The counter is operated at a voltage between the dotted lines (i.e. on the plateau), so that small variations in supply voltage will not affect the response of the instrument.

The whole arrangement of a counting system is illustrated in schematic form in Fig. 7.9. The main function of this type of equipment is the measurement of radioactive samples of various types. In health physics the samples evaluated in this way include air sample filter papers, smear and water samples. The practical aspects of sample counting are described in more detail in Chapter 9 and a number of special techniques are given in Chapter 15.

7.6.4 Pulse height analyser (PHA)

When using detectors from which the output pulse height depends on the energy of the ionizing particle or photon, it is often of great help to analyse the pulses in order to obtain information about the radiation spectrum. A PHA, or kick-sorter as it is sometimes known, separates

Fig. 7.8   Plateau for G-M counter

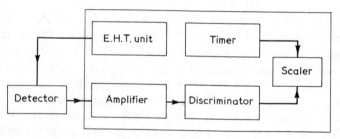

Fig. 7.9   Counting equipment – schematic diagram

the pulses into a large number of channels depending on the pulse
height. Thus if the maximum pulse height in a system is 10 V and 100
channels are available, the pulses can be segregated into channels
0·1 V wide. Any pulse smaller than 0·1 V would go into channel 1,
pulses of 0·1 to 0·2 V into channel 2, and so on up to pulses of 9·9 to
10 V which would go into channel 100. The number of pulses going
into each channel is recorded and presented in a visual display, often
on a cathode ray tube in such a way as to give a visual picture of the
radiation spectrum. Figure 7.10 shows a ^{60}Co γ-ray spectrum as
registered by a NaI crystal.

The two gamma rays of ^{60}Co really have very precise energies but
for various reasons they are smeared out by the NaI detector to give
the two rather broad peaks shown. This has the disadvantage that if a

Fig. 7.10 ^{60}Co γ-ray spectrum from NaI crystal

sample contains a mixture of radionuclides, the peaks may overlap to some extent, making it difficult to resolve the different energies. Germanium detectors offer advantages in this respect since they give very sharply defined lines and permit precise identification of the gamma spectrum and hence the mixture of radionuclides. In order to take advantage of the high resolution of these detectors, modern PHAs have several thousands of channels. A disadvantage of germanium detectors is that they must be maintained at very low temperatures by means of liquid nitrogen cryostats.

7.6.5 Ratemeter

If a detector operating in pulse mode is used in portable equipment it is not usually practicable nor desirable to use a counting system. A more convenient method of presentation is a ratemeter which accepts pulses and indicates on a conventional meter a reading related to the pulse rate.

A basic ratemeter circuit is shown in Fig. 7.11. Each pulse feeds a charge into the capacitor C, which then slowly discharges through the resistance R, giving a reading on the meter. The reading on the meter constantly fluctuates because the arrival of a new pulse causes a sudden increase in reading followed by a slow decrease until the next pulse arrives. The degree of fluctuation can be reduced by increasing the time constant of the circuit, which means increasing the value of C or R or both of them. The time constant in seconds is obtained by multiplying C (farads) by R (ohms). Although an increase of time constant smooths out the meter reading it also slows down the response of the circuit to sudden changes in pulse rate. After a change in pulse rate it takes about four time constants for the ratemeter to

Fig. 7.11   Basic ratemeter circuit

show the new reading. For portable monitoring equipment the time constant should not normally exceed 3 seconds or so.

SUMMARY OF CHAPTER

Ionization in gases: Ion chamber system – current output. Proportional counter – gas amplification – pulse output. Geiger-Müller counter – avalanche effect – pulse output.

Solid state detectors: Ionization, excitation and electron trapping. Conductivity detectors, may be used in pulse or current mode. Scintillation counters – detect light pulses from scintillator using a photomultiplier tube. Thermoluminescence – energy stored until material is heated, then light is emitted. Provides rapid read-out.

Photographic effect: Blackening of film, dose – density curve. Provides permanent dose record.

Activation effect: Measures activation caused by neutrons.

Circuits: d.c. amplifier – measures low currents. Pulse circuits. Pulse amplifier, discriminator, scaler. Pulse height analysers. Ratemeter – registers pulse rate on a meter.

REVISION QUESTIONS

1. Describe the operation of a Geiger-Müller counter.

2. Which class of solid state detector is suitable for measuring a person's accumulated radiation dose? How does the detector function?

3. Write a short description of the photographic effect and discuss its advantages and disadvantages in personal dosimetry.

4. Draw a schematic diagram of a counting system and describe briefly the function of the various circuits.

The external radiation hazard

8.1 SOURCE OF THE HAZARD

The *external radiation hazard* arises from sources of radiation outside the body. When radioactive material actually gets inside the body it gives rise to an *internal radiation hazard*, which requires quite different methods of control. The internal radiation hazard is discussed in Chapter 9.

Alpha radiation is not normally regarded as an external radiation hazard as it cannot penetrate the outer layers of the skin. The hazard may be due to beta, X-, gamma or neutron radiation, all of which can penetrate to the sensitive organs of the body. The external hazard is controlled by applying the three principles: *time*, *distance* and *shielding*.

As elsewhere in the book, this chapter is written primarily in terms of SI units, notably Bq and Sv. However, recognizing the importance of the chapter and the fact that many establishments may adhere to the old units for some time, some of the examples are repeated in terms of rem and Ci in Table 8.1.

8.2 TIME

The dose accumulated by a person working in an area having a particular dose rate is directly proportional to the amount of time he or she spends in the area. His or her dose can thus be controlled by limiting the time spent in the area:

$$\text{Dose} = \text{dose rate} \times \text{time}$$

Example 1 The annual dose limit for category A workers is 50 mSv per year which, assuming a 50 week working year, corresponds to 1 mSv, or 1000 μSv per week. How many hours could a worker spend each week in an area in which the dose rate is 50 μSv/h?

$$\text{Dose} = \text{dose rate} \times \text{time}$$
$$1000 = 50 \times t$$
$$\therefore t = 20 \text{ h}$$

Example 2 If a category A worker has to spend a full 40 h working week in a particular area, what is the maximum dose rate which can be allowed?

$$\text{Dose} = \text{dose rate} \times \text{time}$$
$$1000 = \text{dose rate} \times 40$$
$$\therefore \text{dose rate} = 25 \ \mu\text{Sv/h}$$

Example 3 What would be the maximum dose rate allowed in an area occupied 40 hours per week by category B workers?

$$\text{Dose limit} = 15 \text{ mSv per year} = 15\,000 \ \mu\text{Sv/y}$$
$$\equiv \frac{15\,000}{50} = 300 \ \mu\text{Sv/week}$$
$$\text{dose} = \text{dose rate} \times \text{time}$$
$$300 = \text{dose rate} \times 40$$
$$\therefore \text{dose rate} = \frac{300}{40} = 7 \cdot 5 \ \mu\text{Sv/h}$$

Example 4 The dose limit for individual members of the public is 5 mSv per year. What is the maximum dose rate permitted in an area which could be continuously occupied (i.e. 168 hours per week) by members of the public? (Answer: $0 \cdot 6 \ \mu$Sv/h.)

From these examples it can be seen that the dose rates which are of particular interest and which are commonly encountered in and around facilities such as nuclear reactors are in the range from about 1 μSv/h up to a few tens of μSv/h. In some areas much higher dose rates may be encountered and, as the time to be spent in an area is often dictated by the work to be done, some means of reducing the dose rate to personnel must be employed. The available methods are – increase the distance between the worker and the source of

radiation, or introduce some shielding material between the worker
and the radiation source.

8.3   DISTANCE

Consider a point source of radiation which is emitting uniformly in all
directions. It was shown in Chapter 3 that the flux at a distance r from
a point source is inversely proportional to the square of the distance r.
Since the radiation dose rate is directly related to flux it follows that
the dose rate also obeys the inverse square law. It should be noted that
this is only strictly true for a point source, a point detector and
negligible absorption of radiation between source and detector. The
inverse square law may be written:

$$D \propto 1/r^2 \quad \text{or} \quad D = k/r^2$$
$$\therefore Dr^2 = k$$

where k is a constant for a particular source

$$\therefore D_1 r_1^2 = D_2 r_2^2$$

where D_1 is the dose rate at distance r_1 from the source and D_2 is the
dose rate at distance r_2 from the source.

Example 5   The dose rate at 2 m from a particular gamma source is
400 μSv/h. At what distance will it give a dose rate of 25 μSv/h?

$$D_1 r_1^2 = D_2 r_2^2$$
$$400 \times 2^2 = 25 \times r_2^2$$
$$\therefore r_2^2 = 64$$
$$\text{and } r_2 = 8 \text{ m}$$

It will be noted that doubling the distance from the source reduces
the dose rate to one quarter of its original value, trebling the distance
reduces the dose rate to one ninth, and so on.

Expression for calculating the dose rate from gamma sources

A useful expression for calculating the approximate dose rate from a
gamma source is

$$D = \frac{ME}{6r^2}$$

where D is the dose rate in $\mu Sv/h$, M is the activity of the source in MBq, E is the gamma energy per disintegration, in MeV, and r is the distance from the source in metres.

When applying this expression, care is needed in selecting the correct units. It must be emphasized that in any real situation, protection should be based on measurements of the dose rate.

Example 6 Calculate the approximate dose rate at a distance of 2 m from a 240 MBq cobalt-60 source. Cobalt-60 emits two gamma rays per disintegration, of 1·17 and 1·33 MeV.

$$D = \frac{ME}{6r^2} \; \mu Sv/h$$

$$= \frac{240 \times (1\cdot17 + 1\cdot33)}{6 \times 2^2}$$

$$= \frac{240 \times 2\cdot5}{24}$$

$$= 25 \; \mu Sv/h.$$

Example 7 Calculate the activity of a sodium-22 (^{22}Na) source which gives a dose rate of 64 $\mu Sv/h$ at 1 m. ^{22}Na emits one gamma photon of energy 1·28 MeV per disintegration. (Answer: 300 MBq.)

8.4 SHIELDING

The third method of controlling the external radiation hazard is by means of shielding. Generally, this is the preferred method because it results in intrinsically safe working conditions, whilst reliance on distance or time of exposure may involve continuous administrative control over workers.

The amount of shielding required depends on the type of radiation, the activity of the source and on the dose rate which is acceptable outside the shielding material.

Alpha particles are very easily absorbed. A thin sheet of paper is usually sufficient to stop alpha particles and so they never present a shielding problem.

Beta radiation is more penetrating than alpha radiation. In the energy range which is normally encountered (1–10 MeV) beta radiation requires shielding of up to 10 mm of perspex for complete absorption. The ease with which beta sources may be shielded sometimes leads to the erroneous impression that they are not as dangerous as gamma or neutron sources and large open beta sources are often handled directly. This is an extremely dangerous practice; for instance, the absorbed dose rate at a distance of 3 mm from a beta source of 1 MBq is about 1 Gy/h.

One important problem encountered when shielding against beta radiation concerns the emission of secondary X-rays, which result from the rapid slowing down of the beta particles. This X-radiation, known as *bremsstrahlung*, is discussed more fully in Chapter 12. The fraction of beta energy reappearing as bremsstrahlung is approximately $ZE/3000$ where Z is the atomic number of the absorber and E is the β-energy in MeV. This means that beta shields should be constructed of materials of low mass number (e.g. aluminium or perspex) to reduce the amount of bremsstrahlung emitted.

A beta source emits beta rays with energies covering the complete spectrum from zero up to a characteristic maximum energy, E_{max}. The mean beta energy is, in most cases, about $1/3\ E_{max}$. The penetrating power of β-particles depends on their energy. This fact can be used to estimate the energy of β-rays in order to aid identification of an unknown source. This is discussed in more detail in Chapter 15.

Gamma and X-radiations are attenuated exponentially when they pass through any material. The dose rate due to X- or γ-radiation emerging from a shield can be written as:

$$D_t = D_0 e^{-\mu t}$$

where D_0 is the dose rate without shielding, D_t is the dose rate after passing through a shield of thickness t, and μ is the linear absorption coefficient of the material of the shield.

The linear absorption coefficient μ is a function of the type of material used for the shield and also of the energy of the incident photons. It has the dimensions of (length)$^{-1}$ and is usually expressed in m^{-1} or mm^{-1}.

Half-value layer

The half thickness or half-value layer (HVL) for a particular shielding

Table 8.1 Examples in old units (example numbers as in text)

1 The annual dose limit for category A workers is 5 rem/year which, assuming a 50 week working year, corresponds to 0.1 rem, or 100 mrem per week. How many hours could a worker spend each week in an area in which the dose rate is 5 mrem/h?

$$\text{Dose} = \text{dose rate} \times \text{time}$$
$$100 = 5 \times t$$
$$\therefore t = 20 \text{ h}$$

2 If a category A worker has to spend a full 40 h week in a particular area, what is the maximum dose rate which can be allowed?

$$\text{Dose} = \text{dose rate} \times \text{time}$$
$$100 = \text{dose rate} \times 40$$
$$\therefore \text{ dose rate} = 2 \cdot 5 \text{ mrem/h}$$

3 What would be the maximum dose rate allowed in an area occupied 40 hours per week by category B workers?

$$\text{Dose limit} = 1 \cdot 5 \text{ rem/y}$$
$$= \frac{1500}{50} = 30 \text{ mrem/week}$$
$$= \frac{30}{40} = 0 \cdot 75 \text{ mrem/h}$$

4 The dose limit for individual members of the public is $0 \cdot 5$ rem per year. What is the maximum dose rate permitted in an area which could be continuously occupied (i.e. 168 h/week) by members of the public? (Answer: $0 \cdot 06$ mrem/h.)

5 The dose rate at 2 m from a particular gamma source is 40 mrem/h. At what distance will it give a dose rate of $2 \cdot 5$ mrem/h?

$$D_1 r_1^2 = D_2 r_2^2$$
$$40 \times 2^2 = 2 \cdot 5 \times r_2^2$$
$$\therefore r_2 = \sqrt{64} = 8 \text{ m}$$

8 The dose rate close to a valve is 16 mrem/h due to cobalt-60 inside the valve. How much lead shielding must be placed around the valve to reduce the dose rate to 1 mrem/h? HVL of lead for ^{60}Co gamma radiation is $12 \cdot 5$ mm.

It is required to reduce the dose rate from 16 mrem/h to 1 mrem/h, i.e. by a factor of 16. This requires 4 HVL, $(2 \times 2 \times 2 \times 2 = 16)$ therefore $4 \times 12 \cdot 5$ mm of lead are required, i.e. 50 mm.

9 A certain cobalt-60 source gives a dose rate of 40 mrem/h at 1 m. At what distance from the source must a barrier be placed if the dose rate at the barrier must not exceed $2 \cdot 5$ mrem/h? What thickness of lead would give the same protection at the original distance? (HVL of lead for ^{60}Co gamma radiation is $12 \cdot 5$ mm) (Answer: 4 m; 50 mm of lead.)

material is the thickness required to reduce the intensity to one half its incident value. Writing the half-value layer as $t_{1/2}$, the previous equation becomes:

$$\frac{D_t}{D_0} = 0\cdot5 = \exp(-\mu t_{1/2})$$

Taking logs to the base e:

$$\log_e 0\cdot5 = -\mu t_{1/2}$$
$$\therefore -0\cdot693 = -\mu t_{1/2}$$
$$\therefore t_{1/2} = \frac{0\cdot693}{\mu}$$

The concept of half-value layer is very useful in doing rapid, approximate shielding calculations. One HVL reduces the intensity to one half, two HVLs reduce the intensity to one quarter, three HVLs to one eighth and so on, as illustrated in Fig. 8.1.

Fig. 8.1   Variation of gamma dose rate with absorber thickness

The value of μ, and hence $t_{1/2}$, depends on the material of the medium and on the radiation energy.

Another value sometimes used in shielding work is the *tenth-value layer*, $t_{1/10}$. By a calculation similar to that carried out above it can be shown that:

$$t_{1/10} = \frac{\log_e(10)}{\mu} = \frac{2 \cdot 303}{\mu}$$

Some typical values of $t_{1/2}$ and $t_{1/10}$ for lead and water are given in Table 8.2.

Table 8.2 Approximate values of $t_{1/2}$ and $t_{1/10}$

γ-radiation energy (MeV)	Millimetres of lead		Millimetres of water	
	$t_{1/2}$	$t_{1/10}$	$t_{1/2}$	$t_{1/10}$
0·5	4	12·5	150	500
1·0	11	35	190	625
1·5	15	50	200	700
2·0	19	60	225	750

Example 8 The dose rate close to a valve is 160 μSv/h. If this is due to cobalt-60 inside the valve, how much lead shielding must be placed around the valve to reduce the dose rate to 10 μSv/h? HVL of lead for ^{60}Co gamma radiation is 12·5 mm.

It is required to reduce the dose rate from 160 μSv/h to 10 μSv/h, i.e. by a factor of 16. To do this will require 4 half-value layers of lead $(2 \times 2 \times 2 \times 2 = 16)$, therefore $4 \times 12 \cdot 5$ mm of lead are required, i.e. 50 m.

Example 9 A certain cobalt-60 source gives a dose rate of 400 μSv/h at 1 m. At what distance from the source must a barrier be placed if the dose rate at the barrier must not exceed 25 μSv/h? What thickness of lead would give the same protection at the original distance? (HVL of lead for ^{60}Co gamma radiation is 12·5 mm.)

(Answer: 4 m; 50 mm of lead.)

Neutron shielding is complicated by the very wide range of energies encountered. The most important reactions are:

(*a*) *Elastic scatter*, in which the neutron collides with the target nucleus and 'bounces' off it in a manner similar to the collision of two billiard balls. During the collision, the neutron loses some of its initial energy and this energy is transferred to the target nucleus. All of this transferred energy appears as kinetic energy of the target nucleus. Light elements are best for slowing down neutrons by elastic scatter and so materials with a high hydrogen content (such as paraffin, water, concrete) are used.

(*b*) *Inelastic scatter*; in this process the incoming neutrons impart some of their energy to the scattering material and excite the target nuclei. These target nuclei usually emit gamma radiation later when they return to their ground state. The inelastic scatter process is most important for heavy nuclei.

(*c*) *Neutron capture* reactions of many kinds; in these reactions neutrons are captured by nuclei which then de-excite by emitting another particle or photon. One very important neutron capture reaction is:

$$^{10}B(n, \alpha)^7Li$$

The importance of this reaction, from a shielding point of view, lies in the fact that the emitted particle (an α) is very easily absorbed. Thus the incorporation of boron-10 in shields means that neutrons are absorbed and the resulting α-particles cause no further shielding problems.

Unfortunately, the most common neutron capture reactions lead to the emission of penetrating gamma radiation, for example

$$^{58}Fe(n, \gamma)^{59}Fe$$

Capture gamma radiation is usually a limitation in shield design and a material of high atomic number is often incorporated to absorb these capture gammas.

The neutron reactions are illustrated schematically in Fig. 8.2.

8.5 NEUTRON SOURCES

Apart from the production of neutrons by fission (see Chapter 10), various methods are available for producing relatively small neutron sources. The most common types depend on the reaction:

$$^9Be(\alpha, n)^{12}C$$

(a) Elastic scattering

(b) Inelastic scattering

(c) Neutron capture

Fig. 8.2 The three main neutron reactions

the α-particles being obtained from nuclides such as ^{241}Am or ^{226}Ra. For ^{241}Am/^{9}Be sources, the source strength is about 70 neutrons/ (s MBq) of ^{241}Am. The spectrum of neutrons emitted from an alpha-beryllium source is not mono-energetic but is highly peaked at energies between 3 and 6 MeV, in other words these neutron sources produce mainly fast neutrons.

Another reaction used to produce neutrons is the photoneutron process, that is the (γ, n) reaction. The most common type of photo-neutron source consists of a mixture of equal volumes of antimony and beryllium, in which high energy γ-rays from antimony-124 inter-act with beryllium nuclei causing the ejection of neutrons. It is worth noting that the neutrons produced by the (γ, n) process are, for most practical purposes, monoenergetic.

To calculate the flux at a distance r from a source of strength Q, the following expression is used (see Chapter 3):

$$\Phi = \frac{Q}{4\pi r^2}$$

Flux to dose equivalent rate conversion

In Fig. 8.3 the neutron flux required to give a dose equivalent rate of 25 μSv/h is plotted against neutron energy. From this graph it is seen that for fast neutrons a flux of 10^5n/(m^{2}s)$\equiv$25 μSv/h while for slow neutrons a flux of $\sim 3 \times 10^6$n/(m^{2}s)$\equiv$25 μSv/h. These two approximate values for fast and slow neutrons are very useful in converting calculated neutron fluxes to dose equivalent rates.

Example 10 Calculate the dose equivalent rate at 1 m from a 0·1 TBq americium–beryllium source. 1 TBq of ^{241}Am–Be emits 7×10^7 n/s. (Answer: 140 μSv/h.)

8.6 PERSONAL DOSE CONTROL

Routine control of personal dose is based on a system of area classification. Various systems and terminologies are in use, but there

Fig. 8.3 Relationship between neutron flux and dose equivalent rate (from ICRP data)

is a trend towards that recommended by ICRP. The basic objective is to segregate areas according to the radiological hazard. In areas where the exposure is unlikely to exceed one tenth of the dose limit for exposed workers, i.e. 5 mSv per year, no special arrangements are necessary. Where workers could exceed this level of exposure, but are unlikely to exceed three-tenths of the dose limit, i.e. 15 mSv per year, ICRP suggests that this should be termed a *supervised area*. This would correspond to working condition B defined by ICRP and discussed in Chapter 6. Areas in which exposure could exceed three-tenths of the dose limit, i.e. in which working condition A applies, are called *controlled areas*. Within these controlled areas there may be regions where further demarcation is required to avoid over-exposure. In some establishments these are called *restricted areas*.

Thus a typical system of classification might consider four types of area:

(a) *Uncontrolled areas*, in which the dose rate does not exceed 2·5 μSv/h. Personnel can work for 40 hours per week and 50 weeks per year without exceeding 5 mSv per year.

(b) *Supervised areas*, in which the dose rate does not generally exceed 7·5 μSv/h and hence, in which personnel will not exceed three-tenths of the dose limit. As implied by the name, these areas are subject to some form of supervision, and personnel working regularly in such areas could be subject to routine personal monitoring.

(c) *Controlled areas*, in which the dose rate exceeds 7·5 μSv/h. Personnel working regularly in controlled areas are designated as category A workers and are subject to medical supervision and routine personal monitoring.

(d) *Restricted areas*, in which the dose rate exceeds 25 μSv/h. Access to these areas would be subject to special precautions, such as limitation of access time and the use of protective equipment and monitoring devices.

When operating a system of area classification it is necessary to survey the area regularly in order to confirm that the classification of the area is correct and that adequate precautions are being taken. In controlled and restricted areas, film badges or thermoluminescent dosimeters (TLD) must be worn to measure the accumulated dose to the worker. In addition a direct reading dosemeter such as a quartz fibre electrometer (or electroscope) (QFE) is often worn to give on-the-spot control.

8.7   RADIATION SURVEY MONITORING

8.7.1   Radiation survey monitoring is carried out:

(a) During commissioning of a facility to test the adequacy of the shielding and to show that the radiation levels are satisfactory.
(b) Routinely, during operation, to determine the working radiation levels in order to control accumulated dose.

The ideal radiation survey monitor should be capable of monitoring all forms of penetrating radiation, it should be portable, easy to use and indicate dose equivalent rate. In practice it is not possible to design a single instrument to fulfil all these requirements and so different instruments have been developed for different types of radiation.

8.7.2   X- and γ-monitors

One type of radiation monitor measures X- and γ-radiation and sometimes has a facility to permit an indication (usually not very accurate) of beta radiation. The actual method of detection depends on the sensitivity required. Ion chambers can only be used down to levels of a few tens of μSv/h; below this level the size of chamber required is too large for portable instruments. Increased sensitivity is obtained by using a Geiger-Müller tube or a scintillation detector with a circuit which measures the pulse rate.

The energy response of instruments measuring X- or γ-dose rate is important. In Fig. 8.4 typical response curves are shown for the ion chamber, G-M tube and scintillation detector. It is seen that the ion chamber has a relatively flat response over the energy region 0·3 to 10 MeV while the response curves for the G-M tube and the scintillator tend to peak markedly at low energies. Survey monitors are often calibrated using a radium-226 source, which has an effective photon energy of 0·8 MeV. If the instrument is then used to measure radiation of different photon energy it may seriously underestimate or overestimate the dose rate. Generally, compensating devices are incorporated into instruments using scintillation or G-M detectors to give a relatively uniform response from about 0·1 to 3 MeV.

Fig. 8.4   Energy response curves of various detectors

8.7.3   Neutron monitors

Neutrons, being uncharged, do not ionize directly and so some indirect means has to be used to produce ionization. Fast neutrons are detected by causing them to interact first with a material containing a large proportion of hydrogen atoms. The neutrons 'knock-on' protons in the hydrogenous material and the ionization caused by these protons can be detected. Fast neutron monitors often use a proportional counter with some hydrogenous material, such as polythene, incorporated in its volume. These instruments have a very low sensitivity and it is difficult to detect levels below about 50 μSv/h.

The most common reaction used to detect thermal neutrons is:

$$^{10}B(n, \alpha)^{7}Li$$

Boron-10 has a large cross section for thermal neutron capture and the emitted α-particles cause ionization which may then be detected. The most common thermal neutron monitors employ either an ion chamber lined with a thin layer of boron, or a proportional counter filled with boron trifluoride (BF_3) gas.

The response of instruments using the boron reaction falls off rapidly above energies of a few electronvolts, whereas the instruments using the proton recoil reaction only start to operate at energies above 100 000 electronvolts (0·1 MeV). For many years, in the absence of an instrument to fill this gap it was assumed that the so-called intermediate energy neutrons made an insignificant contribution to total neutron dose. It is now known that in many situations around nuclear reactors intermediate energy neutrons do make an appreciable contribution to neutron dose and instruments have been developed to measure them.

The emphasis is now on *neutron rem meters*, i.e. on instruments which measure tissue dose over a very wide range of energies from thermal up to about 15 MeV. The rem meter illustrated in Fig. 8.6 has a cylinder of polythene which slows down fast neutrons by elastic collisions. A series of cadmium filters are arranged inside the polythene sphere to give the correct energy response function. The thermal neutrons are detected in a proportional counter filled with helium gas. The capture of a neutron results in the emission of a proton:

$$^3He(n, p)^3H$$

where ^3H is an isotope of hydrogen called *tritium*. Once again, ionization is caused by the proton.

This system is now generally preferred to systems using the boron-10 reaction because it is less sensitive to γ-radiation.

8.8 PERSONNEL MONITORING EQUIPMENT

8.8.1 Personal dosimetry

Radiation survey monitoring is used to define radiation levels at various points in a laboratory or around a reactor. It is not an accurate method of assessing the accumulated dose received by workers in these areas because:

(a) it is quite likely that the dose rates will vary considerably with time, depending on the operations being carried out;
(b) the workers will usually move around from one radiation level to another during the course of their work.

To overcome these difficulties, it is normal practice for people working in radiation areas to wear a *personal dosemeter*. This is a device which measures the dose accumulated by the wearer and there are several types of personal dosemeter in common use.

Fig. 8.5 Portable gamma survey meter (courtesy of Nuclear Enterprises Ltd)

Fig. 8.6 Neutron dose equivalent meter (courtesy of Nuclear Enterprises Ltd)

8.8.2   The film badge

This is the most common personal dosemeter and consists of a piece
of photographic film in a special holder. The film badge is one of the
accepted methods of monitoring whole-body radiation dose for the
purpose of record keeping. The developed films may be stored and
could if necessary, be rescrutinized later.

The film badge in general use in the United Kingdom consists of a
Kodak Personal Monitoring Type 2 film in a special holder. The
Kodak Type 2 film is a double emulsion type having a fast emulsion
on one side and slow emulsion on the other. The degree of blackening
on the developed film is determined using a densitometer and then
related by calibration to the radiation exposure of the film. The use of
two emulsions permits measurement over a wide range of dose. The
fast emulsion enables γ-ray doses in the range 50 μSv to about 50 mSv
to be measured. If a dose in excess of about 50 mSv has been received,
part of the fast emulsion is stripped from the film and the slow
emulsion then permits measurements up to 10 Sv.

The holder is illustrated in Fig. 8.7. It incorporates several filters so
that β-, γ-, X- and thermal neutron doses can be measured. Basically
beta dose is measured in the open window area, gamma under the
lead and thermal neutron by taking the difference between the (lead
+ cadmium) and the (lead + tin) filters. Thermal neutrons interact
with cadmium via an (n, γ) reaction and the resulting gamma rays give
additional blackening under the cadmium filter. The 300 mg/cm^2 and
50 mg/cm^2 filters permit energy corrections to be applied to the β-
and low-energy γ-doses. The strip of lead around the edge of the
holder is to minimize edge effects at the boundary of the various
filters. The indium foil is an 'exposure indicator' for criticality accidents. A dose above 10 mSv (1 rem) of thermal neutrons will activate
the indium sufficiently to permit measurement using a Geiger-Müller
tube.

8.8.3   Thermoluminescent dosemeters

These materials offer an accurate and stable means of measuring dose
over the short and long term and find applications both as whole-
body and extremity monitors. The action of ionizing radiation on
thermoluminescent materials and the method of reading has been
described in Chapter 7. One of the disadvantages of this technique is

Filter Types
1. Window
2. 50 mg/cm² plastics
3. 300 mg/cm² plastics
4. 0·004″ Dural
5. 0·028″ Cd 0·012″ Pb
6. 0·028″ Sn 0·012″ Pb
7. 0·012″ Pb edge shielding
8. 0·4 gm of indium

Films exposed in the holder to various types of radiation

Beta, thallium 204,
$E_{max} = 0·78$ MeV

Beta, yttrium 90,
$E_{max} = 2·25$ MeV

X-rays, 20 keV

X-rays, 150 keV

Gamma rays,
cobalt 60

Thermal neutrons

Fig. 8.7   The film badge. (After Heard and Jones, courtesy of UKAEA) that the process of reading the dose destroys the information, so that, unlike the film badge, the dosemeter can be read only once. The TLD system also gives less information about the quality of the radiation.

Two materials currently in use are lithium fluoride and calcium fluoride. The latter is very sensitive but has a poor energy response. Lithium fluoride is less sensitive but its energy response is excellent. In a practical dosemeter system the thermoluminescent material is usually in the form of a powder or a thin disc.

The majority of establishments are now using TLD systems as the primary method of personal monitoring. This is because they are particularly suitable for automatic linking to computerized dose recording systems. Even in establishments in which the film badge is the main method of personal monitoring, TLD systems are used to provide a convenient method of short-term dose control.

8.8.4   Fast neutron dosemeter

The fast neutron dosemeter consists of a nuclear emulsion film sealed in a moisture proof sachet and worn in a holder similar to a film badge. The holder contains lead, boron and plastic filters to screen the emulsion from body-scattered neutron radiation. The fast neutrons interact with the base material of the film and cause recoil protons to be ejected. These protons create ionization tracks in the emulsion which show up when the film is developed. The tracks are counted under a microscope and the number of tracks per square centimetre is a measure of the neutron dose.

Track plates have a range of 1 mSv to 1 Sv. Their chief drawbacks are that they are tedious and expensive to evaluate and that they have a threshold energy of about 0·5 MeV for detection of neutrons. In addition, they are sensitive to gamma radiation and a dose of about 100 mSv of γ makes the counting of tracks impossible.

8.8.5   Criticality locket

A criticality locket is worn in addition to the film badge or TLD whenever fissile material is handled (e.g. fuel element manufacturing and processing plants, fuel element cooling ponds and reactors). It is designed to measure the very high neutron doses which would be experienced during a criticality accident. The criticality locket con- tains components which are activated by neutrons of different energies. The reactions utilized in the dosemeter are:

$$^{32}S(n, p)^{32}P \qquad (S - sulphur, P - phosphorus)$$
$$^{197}Au(n, \gamma)^{198}Au \qquad (Au - gold)$$
$$^{115}In(n, n)^{115}In \qquad (In - indium)$$

All of the activated components are beta emitters and can be counted in a beta castle. From the counts obtained the fast, intermediate and thermal neutron dose can be estimated.

8.8.6 Quartz fibre electrometer

The quartz fibre electrometer (QFE) is a pocket dosemeter, about the size of a large pen, which gives continuous visual indication of accumulated gamma dose. It consists of a quartz fibre assembly in a small ion chamber, a spring loaded plunger for charging the electrometer, a microscope lens through which to view the quartz fibre and a scale calibrated in roentgens. The electrometer is charged by applying a voltage to the centre electrode which causes the quartz fibre to deflect. When it is exposed to gamma radiation the air inside the chamber is ionized and the charge on the electrode reduced. Consequently the deflection of the quartz fibre is decreased and the change in deflection indicates the dose. Quartz fibre electrometers are available with sensitivities ranging from full scale deflections of 1 mSv to 100 Sv.

8.8.7 Electronic dosemeters

Various types of electronic personal dosemeters are available, usually based on miniature G-M tubes. They provide a direct display of either dose rate or accumulated dose. One type is intended as a replacement for the QFE and has a similar pen-like shape.

8.9 RADIATION RECORDS

The purpose of personal dosimetry is to ensure that workers exposed to ionizing radiation are kept within the dose equivalent limits specified in the appropriate legislation or code of practice. In most countries, personal monitoring devices other than those for short-term dose control must be provided by an approved laboratory.

An approved dosimetry laboratory has the following duties:

(a) Issue film badges, thermoluminescent dosemeters and any other personal dosemeters (such as fast neutron track plates) which may be necessary.
(b) Process and assess these on return.
(c) Issue dose reports and maintain dose records.

Some of the larger employers of workers who wear personal dosimetry devices have established their own processing laboratories. Other users rely on specialist organizations such as, in the UK, the National Radiological Protection Board.

For legal purposes both film badges and TLD are approved methods of personal dosimetry in the UK. They are normally processed once a month and the results are recorded in the worker's personal dose record. In addition to a monthly dose report, a quarterly summary is normally issued in respect of each employee which summarizes the total radiation dose accumulated over the calendar quarter, year and working life.

SUMMARY OF CHAPTER

External radiation hazard: arises from radioactive materials outside the body.
Control of external hazard: time, distance and shielding.
Time: Dose = dose rate × time.
Distance: Inverse square law $D_1 r_1^2 = D_2 r_2^2$.
    Dose rate from a gamma source

$$D = \frac{ME}{6r^2} \, \mu Sv/h$$

(M in MBq, E in MeV, r in metres).
Shielding: *Alpha* particles very easily absorbed.
        Beta radiation. Use low Z materials to reduce bremsstrahlung.
        Gamma radiation is attenuated exponentially,
        $D_t = D_0 e^{-\mu t}$. Half-value layer $t_{1/2} = 0.693/\mu$.
        Neutron shielding – elastic scatter, inelastic scatter, and neutron capture.

Neutron sources: depend on either (α, n) or (γ, n) reactions.

Area classification: e.g. uncontrolled, supervised, controlled, and restricted areas.

Radiation survey monitors: *X- and γ-monitors* use ion chambers, Geiger-Müller tubes or scintillation detectors.

Neutron Monitors: fast neutrons detected by knock-on protons; thermal neutrons usually via the reaction ^{10}B(n, α)^{7}Li;

neutron rem meters cover the energy range from thermal up to about 15 MeV, use the reaction ^{3}He(n, p)^{3}H to give good gamma rejection.

Personal dosemeters: *Film badge*, special film with two emulsions to cover range from 50 μSv to 10 Sv, in a holder with filters which allow measurement of β, γ, X and thermal neutron dose.

Thermoluminescent dosemeters store the radiation energy which can later be released by heating. The light output is measured using a photomultiplier tube, the electrical output of which is a measure of the radiation dose.

Fast neutron track plate, special film in a holder, fast neutrons eject recoil protons which cause developable tracks in the emulsion. Main disadvantage – track plates are expensive to evaluate.

Criticality locket, worn when handling fissile material; the various components are activated by neutrons of different energy and can be counted in a β-castle.

Quartz fibre electrometer, pocket ionization chamber which gives continuous indication of accumulated gamma dose.

REVISION QUESTIONS

1. What are the three methods by which the external radiation hazard is controlled?

2. To carry out a certain process a worker has to work in an area which has an average dose rate of 100 μSv/h. How many hours per week can he or she work in this area if 1 mSv/week is not to be exceeded? To what level must the dose rate be reduced to allow work in the area for 40 hours each week?

3. Calculate the dose equivalent rate at a distance of 1 metre from a 540 MBq cobalt-60 source. At what distance will the dose equivalent rate be 25 μSv/h?

4. The dose rate at a distance of 1 m from a certain gamma source is 900 μSv/h. At what distance from the source is the dose rate 25 μSv/h?

5. Calculate the approximate dose rate at a distance of 2 m from a 3000 MBq gamma source which emits one 1·6 MeV gamma photon per distintegration.

6. Discuss the problem of detecting neutrons and show how it is overcome in (*a*) fast neutron monitors, (*b*) thermal neutron monitors, (*c*) neutron rem meters.

7. List the main types of personal dosemeter and discuss the relative advantages and disadvantages of the film badge and the TLD.

9

The internal radiation hazard

9.1 UNCONTAINED RADIOACTIVITY

When a radioactive material is enclosed inside some form of sealed container it may give rise to an external radiation hazard to personnel working in its vicinity. On the other hand, when radioactive material is not contained in any way it constitutes a potential *internal radiation hazard*. Radioactive material which is inadvertently released is known as *contamination*.

Quite small quantities of radioactive material which represent an insignificant external hazard can give rise to appreciable dose rates if they come into contact with, or get inside the body. Once the radioactive substance is taken into the body it will continue to irradiate the body until either the radioactivity has decayed or until the body has excreted the substance. The rate of decay of the radioactivity depends on its half-life, which can vary from a small fraction of a second to many thousands of years. The rate of excretion of the substance from the body depends on a number of variables such as the chemical characteristics of the substance, and it may happen in a period of a few days or it may take much longer, perhaps up to many years. Thus when a radioactive substance enters the body it may irradiate it for only a few days or for a much longer period which may extend to many years in the case of certain nuclides.

9.2 ROUTES OF ENTRY

There are four ways in which contamination can present a hazard to the body. These are:

(*a*) direct *inhalation* of airborne contamination,
(*b*) *ingestion*, that is entry through the mouth,
(*c*) entry through the skin, or through a contaminated wound,
(*d*) direct irradiation of the skin.

When contamination is present in the atmosphere it will be breathed into the lungs and a certain fraction of it will pass into the bloodstream. Another fraction of the inhaled contamination is eliminated from the lungs and swallowed; the remainder is exhaled. The various fractions passed into the bloodstream, swallowed or exhaled depend on many factors such as the chemical and physical form of the contamination, and the physiology of the person involved. Similarly, when contamination is ingested, the amount of it passing through the wall of the digestive tract into the body fluids depends on the nature of the contamination and on the physiological conditions.

There is a wide variation in the physiological characteristics of human beings and, for the purposes of radiological protection, the International Commission on Radiological Protection (ICRP) has defined a *reference man*, some of whose characteristics are listed in Table 9.1. It will be seen, for instance, that reference man breathes about 23 m^3 of air per day and has a total water intake of 3 litres/day. Reference man is an entirely fictitious individual and simply represents an average over the very wide spectrum of human characteristics.

It was pointed out above that the fate of a particular radioactive nuclide inside the body depends on its chemical and physical form. For instance, some elements distribute themselves fairly uniformly and so irradiate the whole body at about the same rate. The majority of elements, however, tend to concentrate in particular organs so that an intake of radioactivity may result in different dose rates to the various organs of the body. Examples of such elements are iodine, which concentrates in the thyroid gland, and plutonium, which concentrates in the lung or bone. The system of dose limitation given by ICRP in Publication 26 allows the doses to individual organs and tissues to be computed using the weighting factor formula described in Chapter 6.

The dose rate to any organ is proportional to the amount of radioactivity in the organ and decreases as the radioactive isotope decays or is excreted. The radioactive decay of an isotope is exponential in character and it is found that the rate of excretion of most substances from the body may also be considered as approximately exponential. This means that an *effective decay constant* can be employed to describe the rate of removal of a radioactive substance from the body, namely (see Fig. 9.1):

$$\lambda_{\text{eff}} = \lambda_r + \lambda_b$$

Table 9.1 Some characteristics of reference man

(a) Organs of reference man

Organ	Mass, m (kg)	Per cent of total body
Total body	70	100
Skeleton	10	14
Muscle	28	40
Fat	13·3	19
Blood	5·5	7·9
Gastrointestinal tract (incl. contents)	2·2	3·1
Thyroid gland	0·02	0·029

(b) Air and water balance

Water intake (litres/day)		Excretion (litres/day)	
Foods	0·7	Urine	1·4
Fluids	1·95	Sweat	0·65
Oxidation	0·35	Insensible	0·85
		Faeces	0·1
Total:	3·0	Total:	3·0

Air balance

Vital capacity of lungs 4·3 litres	
Air inhaled during 8-hour working day	9·6 m^3
Air inhaled during 16 hours not at work	13·2 m^3
	Total: ~ 23 m^3/day

These values are for the adult male. In most cases the values for adult female are lower. For full details of the characteristics of reference man see ICRP Publication 23.

Fig. 9.1   Typical elimination curve of a radionuclide in the body

where

λ_r = radioactive decay constant

λ_b = biological decay constant

Since the decay constant is equal to $\log_e 2/\text{half-life}$, this equation becomes:

$$\frac{1}{T_{\text{eff}}} = \frac{1}{T_r} + \frac{1}{T_b}$$

where

T_{eff} = effective half-life of a radioactive substance in the body

T_r = radioactive half-life of the substance

T_b = biological half-life of the substance

Figure 9.2 illustrates the variation of dose rate with time following an intake of a radionuclide. The initial rise in the curve covers the period during which the nuclide is being transported to the organ of interest. At the peak most of the radionuclide that is destined for the particular organ has reached it and the organ is receiving its maxi-

Fig. 9.2 Variation of dose rate with time following an intake of a radionuclide

mum dose rate. Subsequently the dose rate to the organ decreases approximately exponentially as the radionuclide decays and is excreted. The total dose received by the organ is obtained by evaluating the area under the curve. Thus a given intake of a radionuclide will commit the organ (or organs) at risk from that nuclide to a certain dose, which is known as the *dose commitment*. The dose commitment depends on the initial dose rate and on the removal rate. It is usually assumed that a given intake of a particular radionuclide will result in the same dose commitment whether it is received in a single large intake or a large number of small intakes. A special case of *dose commitment* is called *committed dose* and is the dose accumulated over 50 years, representing the likely maximum lifespan of the individual following the intake.

9.3 ANNUAL LIMITS OF INTAKE

Committee 2 of ICRP, dealing with internal exposure, has established annual limits of intake (ALIs) for a wide range of radionuclides.

The ALI is the amount of a radionuclide (in Bq) which would give a harm commitment to the organs it irradiates equal to that resulting from whole-body irradiation of 50 mSv (5 rem). For example, consider a radionuclide which, when taken into the body, irradiates organs X, Y and Z. Suppose that, for an intake of 1 Bq of the radionuclide, the committed dose equivalent to each of these organs is H_x, H_y and H_z, respectively. If the weighting factors for organs X, Y and Z are w_x, w_y and w_z then the effective dose equivalent from an intake of 1 Bq is:

$$H = w_x H_x + w_y H_y + w_z H_z$$

The ALI is the quantity such that

$$\sum_T w_T H_T = 50 \text{ mSv}$$

The ALI is then given by

$$\frac{50}{w_x H_x + w_y H_y + w_z H_z} \quad \text{(in Bq)}$$

(*Note*: For conversion factors relating Bq and Ci refer to Chapter 2).

The annual limit of intake is set such that the sum of risks of stochastic effects from the doses to the various body organs in the 50 years following the intake will not exceed the risk of stochastic effects associated with uniform irradiation of the whole body to the annual dose limit of 50 mSv (5 rem). However, there are overriding requirements, based on non-stochastic effects, that the dose to any single organ must not exceed 0·5 Sv (50 rem), or 0·15 Sv (15 rem) to the lens of the eye, in a year. These non-stochastic limits are more restrictive on intake than the stochastic limits for about 20% of radionuclides.

In Table 9.2, ALI values are given for some important radionuclides, for inhalation and ingestion. Since the transfer of material from the lung or from the gut is influenced by the chemical form, it is necessary to specify different values of ALI for certain chemical compounds of some nuclides. Where non-stochastic effects determine the limits, the organ or tissue at risk is shown with the value of ALI. Thus the ALIs for sodium-22 and caesium-137 are determined by the stochastic dose limits, while those for iodine-131 and plutonium-239 are determined by the non-stochastic limits.

In controlling and assessing the total dose which a person receives in a year, both the external and internal doses must be considered. Following ICRP Publication 26 the following formula may be used to combine the effects of external and internal exposures that are

Table 9.2 Some values of ALI

Nuclide	Compound	Inhalation ALI (Bq)	Ingestion ALI (Bq)
Sodium-22	All compounds	2×10^7	1×10^7
Iodine-131	All compounds	2×10^6 (thyroid)	1×10^6 (thyroid)
Caesium-137	All compounds	6×10^6	4×10^6
Plutonium-239	Plutonium dioxide	5×10^2 (bone)	
	All other compounds	2×10^2 (bone)	
	Oxides and hydroxides		2×10^6 (bone surface)
	All other compounds		2×10^4 (bone surface)

Note: For conversion factors relating Bq and Ci refer to Chapter 2.

received together, to ensure that the recommended dose limitation for stochastic effects will not be exceeded:

$$\frac{H_{wb}}{H_{wb,L}} + \sum_j \frac{I_j}{I_{j,L}} \leq 1$$

where H_{wb} is the dose equivalent received in the year from external exposure, $H_{wb,L}$ is the annual dose equivalent limit, I_j is the intake of radionuclide j during the year, and $I_{j,L}$ is the annual limit of intake for radionuclide j.

Example 1 During a particular year it is estimated that a worker will be exposed to intakes of 10^6 Bq of sodium-22 (via ingestion) and 10^2 Bq of plutonium-239 dioxide (via inhalation). What is the maximum dose equivalent that he can receive from external exposure during the year if the recommended dose limit is not to be exceeded?

For sodium-22, the ingestion ALI $= 10^7$ Bq and so in this case,

$$\frac{I_j}{I_{j,L}} = \frac{10^6}{10^7} = \frac{1}{10}$$

For plutonium-239 dioxide, the inhalation ALI $= 5 \times 10^2$ Bq and so in this case,

$$\frac{I_j}{I_{j,L}} = \frac{10^2}{5 \times 10^2} = \frac{1}{5}$$

Using the formula

$$\frac{H_{wb}}{H_{wb,L}} + \sum_j \frac{I_j}{I_{j,L}} \leq 1$$

$$\frac{H_{wb}}{50 \text{ mSv}} + \frac{1}{10} + \frac{1}{5} = 1, \text{ in the limit}$$

$$\therefore H_{wb} = (1 - 0.1 - 0.2) \ 50 \text{ mSv}$$
$$= 0.7 \times 50 \text{ mSv}$$
$$= 35 \text{ mSv } (3.5 \text{ rem})$$

Thus the worker would be permitted to receive 35 mSv (3·5 rem) from external exposure in the year.

9.4   CONTROL OF THE CONTAMINATION HAZARD

As with external radiation, the consideration in the control of the radioactive contamination hazard is to limit the dose to the various organs of the body to the permitted level. The hazard is controlled by limiting the airborne concentration and the level of surface contamination.

9.4.1   Derived air concentrations

The derived air concentration (DAC) is the concentration of the radionuclide in the air which would result in a worker receiving an intake of 1 ALI in a year through inhalation. The DAC is obtained by dividing the inhalation ALI of the particular nuclide by the volume of air breathed in the working year. Using the ICRP 'reference man' breathing rate, the DAC may be calculated. The breathing rate is assumed to be 10 m^3 per 8 hour working day.

Thus for a worker employed for 50 weeks per year, 5 days per week, 8 hours per day, the DAC is:

$$DAC = \frac{ALI(\text{inhalation})}{50 \times 5 \times 10} \text{ Bq/m}^3$$

Example 2   Calculate the DAC for plutonium-239 dioxide. ALI (inhalation) for $^{239}PuO_2 = 5 \times 10^2$ Bq.

$$DAC = \frac{5 \times 10^2}{50 \times 5 \times 10} = 2 \times 10^{-1} \text{ Bq/m}^3$$

Values of DAC vary enormously, ranging, for example, from 8×10^{-2} Bq/m^3 ($2 \cdot 2 \times 10^{-12}$ μCi/cm^3) for most compounds of ^{239}Pu up to 10^8 Bq/m^3 ($2 \cdot 7 \times 10^{-3}$ μCi/cm^3) for certain other radionuclides.

9.4.2   Derived limits for surface contamination

The hazards due to surface contamination are controlled by setting *derived limits* (DL) of surface contamination such that:

- (*a*) the amount kicked up does not cause the DAC to be exceeded,
- (*b*) the amount ingested via the hands and mouth does not cause the ALI to be exceeded,
- (*c*) if on the skin (normally of the hands), the skin dose does not exceed the permitted level.

In most cases it is found that the skin dose, or potential skin dose, is the limiting consideration for β-emitters. For α-emitters the limit is normally defined by the possible inhalation hazard.

The Derived Limit depends on the radiotoxicity of the nuclide or mixture of nuclides present; values of typical α-emitters and typical mixtures of β-emitters are shown in Table 9.3. The limit for most α-emitters is a factor of 10 lower than that for nuclides which emit other radiations; this reflects the very high toxicity of most α-emitters. In areas in which the DL is liable to be exceeded special control measures must be exercised.

Table 9.3   Derived limits of surface contamination

Very high toxicity* nuclides including α-emitters	Medium toxicity* nuclides emitting other radiations	Lower toxicity* nuclides
3×10^3 Bq/m^2 (8×10^{-6} μCi/cm^2)	3×10^4 Bq/m^2 (8×10^{-5} μCi/cm^2)	3×10^5–3×10^6 Bq/m^2 (8×10^{-4}–8×10^{-3} μCi/cm^2) depending on toxicity

*For the classification of nuclides according to their radiotoxicity, see Section 9.6.

9.5   ROUTINE CONTROL OF CONTAMINATION

9.5.1   Basic principles

There are three basic principles which can be applied to the control of
radioactive contamination:

(a) *Minimize* as far as possible the amount of activity being
handled.

(b) *Contain* radioactive material, normally at least two levels of
containment are provided.

(c) Follow the *correct procedures* regarding protective clothing,
washing and monitoring facilities, etc.

Figure 9.3 illustrates a typical containment system which might be
applied to a radiochemistry laboratory. The four levels of containment are: the bottle containing the liquid, the splash tray, the fume
hood and finally the barrier at the entrance to the laboratory.

9.5.2   Area classification

As with the external radiation hazard, routine control of contamination is by means of a system of *area classification*. Table 9.4 shows the

Levels of containment:

(i)   Bottle containing radioactive liquid
(ii) Splash tray
(iii) Fume cupboard
(iv) Barrier with contamination monitor

Fig. 9.3   Schematic diagram illustrating four levels of containment

Table 9.4 Control levels for area classification

Type of area	Surface contamination	Airborne contamination
Uncontrolled (non-active)	less than 1 DL	less than 1/10 DAC
Supervised (contamination)	1 to 3 DL	1/10 to 3/10 DAC
Controlled (contamination)	greater than 3 DL	greater than 3/10 DAC

control levels of a typical system. It must be stressed that this table should be regarded as illustrative only. Many establishments have more detailed systems of area classification to cope with the specific contamination hazards that arise from the particular work activity.

Whenever possible contamination should be cleaned up as soon as it occurs. This prevents further spread which makes the eventual decontamination more difficult.

Regular surveys should be made in supervised and controlled areas and in the adjacent uncontrolled areas to ensure that contamination is not spreading beyond the barriers.

9.5.3 Protective clothing

The protective clothing requirements in a contaminated area depend on the nature and amount of the contamination. For low levels of surface contamination an ordinary laboratory coat with overshoes and gloves may be sufficient. When there are substantial levels of airborne contamination it is usually necessary to have a fully-enclosed dry suit and a filter mask or a mask fitted with an air supply. Again, when the contamination is in liquid form it is often necessary to wear a fully enclosed PVC suit with a filter mask or fresh air supply. Some examples of protective clothing are illustrated in Fig. 9.4.

Whatever the standard of protective clothing, the change area and barrier arrangements must be efficient and should have the following facilities:

(a) Cotton overall, cap, gloves and overshoes. Suitable for levels of up to 100 DL if dry.

(b) Impervious clothing and filter respirator.

(c) Fully ventilated suit is much more comfortable and more efficient than equipment in (b).

(d) Crash suit for emergency use, and self contained breathing set.

Fig. 9.4 Protective clothing and equipment. (Crown copyright. Reproduced by permission of the controller HMSO) (*a*) Washhand basin (and possibly a shower) and monitoring instruments (for example, a hand and clothing monitor).

(*b*) Suitable stowage on the non-active side of the barrier for the worker's personal clothing.

(*c*) Conveniently-placed protective clothing ready for use.

(*d*) Containers for used clothing and radioactive waste.

(*e*) Noticeboards at the barrier stating 'no unauthorized entry', the hazards in the area, the clothing to be worn and any other precautions to be taken.

(*f*) Emergency instructions, detailing actions in the event of possible incidents such as criticality, fire, serious personal contamination, should be posted in the area. Consideration must also be given to suitable emergency exits.

Special arrangements are made for laundering clothing worn in contaminated areas and the effluent from laundry facilities is treated as liquid radioactive waste.

9.5.4 House rules and training of personnel

The control of contamination depends on everyone who enters a controlled or restricted area and so all personnel who work in such areas should be given initial and subsequent periodic training in the hazards involved and in the house rules. Some typical house rules for controlled and restricted areas are:

(*a*) No eating, drinking or smoking.

(*b*) No mouth operations (such as pipetting).

(*c*) Any wounds should be covered with waterproof dressing prior to entering the active area. This is most important since open wounds provide a direct route for contamination into the bloodstream.

(*d*) Wounds sustained in the area are to be reported to the person in charge and treated immediately.

(*e*) Ordinary handkerchiefs should not be used in an active area. Disposable tissues should always be available.

(*f*) All items being removed from an active area are subject to Health Physics clearance before being permitted to leave the area. Active items should be suitably labelled. Whenever possible, sets of tools and cleaning gear should be reserved solely for use in active areas and should be clearly marked as 'active'.

9.5.5 The international symbol for radiation

The internationally agreed symbol for ionizing radiation is the trefoil symbol illustrated in Fig. 9.5. This symbol is found on packages containing radioactive sources, at the entrance to areas where there is a significant radiological hazard, etc.

(Black on yellow ground)

Fig. 9.5   The trefoil symbol

9.6   RADIOTOXICITY AND LABORATORY CLASSIFICATIONS

A toxicity classification for radionuclides was recommended by the International Atomic Energy Agency (*IAEA Technical Reports Series No. 15*) to act as a guide to the procedures and facilities required for handling unsealed radioactive substances. In the Euratom Directive on radiation protection (see Chapter 14) the same classification scheme is adopted, in which radionuclides are divided into four groups according to their radiotoxicity:

Group I	high toxicity (e.g. ^{239}Pu, ^{241}Am)
Group II	upper medium (e.g. ^{90}Sr, ^{131}I)
Group III	medium and lower medium (e.g. ^{32}P, ^{65}Zn)
Group IV	low (e.g. ^{129}I, natural uranium)

There are many nuclides in each group, the nuclides mentioned are common examples.

In addition, three classes of laboratory are defined by the IAEA, called classes 1, 2 and 3. A class 1 laboratory is a specially designed facility with elaborate equipment to enable safe handling of high levels of radioactivity. Class 2 laboratories are of a standard comparable to any high quality chemistry laboratory, whereas class 3 includes ordinary laboratories not originally designed for handling toxic materials. The quantities of radionuclides that can be handled with reasonable safety in the various classes of laboratory are shown in Table 9.5.

9.7 DESIGN OF AREAS FOR RADIOACTIVE WORK

A considerable amount of pre-planning must be done before an active laboratory or active area is set up. Apart from the general features of design expected in any good facility, special attention must be paid to ventilation and to surface finishes. Ventilation systems require efficient filtration units to remove particulate activity prior to discharge. In the case of gaseous activity, which would not be removed by filtration, great care must be exercised in the location of outlets to ensure adequate dispersal of discharged activity.

Surfaces in an active area should be smooth and unbroken and be made from materials which are chemically inert, non-absorbent and water repellant. Consideration must also be given to possible decontamination problems which might arise, and so materials must be chosen which are either easily decontaminated or which can be conveniently removed and replaced.

Walls, floors and ceilings

The basic requirement is that the walls, floors and ceiling should have a good clean finish which is free from cracks. From the point of view of cleanliness and ease of cleaning it is desirable to have covings at all angles of walls, ceiling and walls, and floors and walls.

Plastered walls and ceilings can be made non-porous and smooth by the application of gloss paint. An alternative approach is to 'face' the walls with a suitable strippable material, for example, melamine laminate on a resin bonded plywood backing which is butt jointed and the joints sealed. Other finishes which have been used for walls

Table 9.5 Guide to quantities of radionuclides which may be handled

Radionuclide toxicity	Class of laboratory		
	3	2	1
Group I	Up to 370 kBq (10 µCi)	Up to 370 MBq (10 mCi)	Above 370 MBq (10 mCi)
Group II	Up to 3·7 MBq (100 µCi)	Up to 3·7 GBq (100 mCi)	Above 3·7 GBq (100 mCi)
Group III	Up to 37 MBq (1 mCi)	Up to 37 GBq (1 Ci)	Above 37 GBq (1 Ci)
Group IV	Up to 370 MBq (10 mCi)	Up to 370 GBq (10 Ci)	Above 370 GBq (10 Ci)

Note: These figures are for normal, wet chemical operations. Modifying factors may be applied as follows:

Work	Modifying factor
Storage in closed, vented containers	× 100
Simple wet chemistry – low specific activity	× 10
Complex wet, or simple dry operations	× 0·1
Dry and dusty operations	× 0·01

and ceilings include chlorinated rubber-based paint, epoxide resin paint and various other strippable materials.

The most satisfactory floor covering consists of sheet PVC which is stuck down and all joints welded. Such a floor covering should have an integral coved skirting. An alternative covering consists of sheet linoleum which is made water repellant by applying a hard wax followed by soluble wax. In the case of sheet linoleum a separate pre-formed coved skirting has to be used and joints have to be cold welded. PVC or linoleum tiles are not normally recommended because the many joints in the floor make cleaning-up after contamination a difficult operation. Concrete and wood are poor floor materials but their use is sometimes unavoidable. When they have to be used they should be treated with a rubber-based paint to make them water repellant.

Working surfaces

Working surfaces should be finished in hard non-porous materials which have the necessary heat and chemical resisting properties.

The materials most commonly used include:

(a) Melamine resin plastic laminate such as Formica. It should be bonded to the backing material with a resin glue to give the necessary temperature resistance.

(b) Polyvinyl chloride sheet, such as Darvic, which can be welded and is completely self-extinguishing.

(c) Stainless steel is a useful material but there is a tendency to get physical bonding between it and corrosion products. Also, stainless steels are susceptible to attack from certain chemicals, for example, hydrochloric acid.

(d) Glass fibre reinforced resin which can be moulded to shape. It can be treated to make it fire resistant, but it may burn in a fire.

(e) Polypropylene, which can be welded and heat-formed. This material has such a high chemical resistance that it is difficult to find suitable adhesives for it. It is not fire resistant and, once ignited, it will continue to burn.

Glove boxes

These are installed in active laboratories to facilitate the handling of hazardous materials. They consist of a leak-tight enclosure in which

objects or materials may be manipulated through gauntlet gloves
attached to ports in the walls of the box (see Fig. 9.6). Their aim is to
provide containment for materials which are either radioactive, or
chemically toxic, or both. Usually they do not provide any shielding
protection against penetrating radiation and so they are used for
alpha or beta emitters. When gamma-emitting isotopes have to be
handled, a wall of lead bricks is usually constructed between the
operator and the glove box.

Glove boxes are maintained at a pressure slightly below that of the
outside laboratory. This means that air will flow into the glove box
should a leak develop and this will prevent the contamination escaping. Two filters are normally placed on the ventilation system – one
to remove dust from the air being drawn into the glove box and the
second to remove radioactive particles from the air being drawn out
of the box.

Fume cupboards

A fume cupboard is used when relatively low levels of activity (in the
MBq range) are being handled. The material is handled via an
opening at the front through which air is drawn from the laboratory
into the fume cupboard (see Fig. 9.7). This protects the operator from
any leakage of contamination through the opening into the general

Fig. 9.6 Schematic drawing of a glove box

Fig. 9.7 Schematic drawing of a fume cupboard

laboratory. The services usually required in a fume cupboard are water, gas, vacuum and electricity. The controls for these services should be situated outside the fume cupboard to minimize the number of movements through the front opening. It is good practice to ensure that the openings of the fronts of fume cupboards are kept to a minimum to reduce the chance of radioactive contamination entering the general laboratory atmosphere.

9.8 TREATMENT OF CONTAMINATED PERSONNEL

Once a radioisotope has become lodged in the body, very little can be done to increase the rate of elimination (i.e. decrease the biological half-life) of the isotope. This means that every effort must be made to prevent contamination entering the body. To this end it is vital that all personnel should obey the house rules and always wear the correct protective clothing. Even so, contamination incidents are bound to occur and so a knowledge of the correct treatment is vital.

The first action when dealing with a contaminated person is to ascertain whether or not he or she is injured. If there is serious injury then first-aid treatment must be given as quickly as possible. Following any necessary medical treatment, the next actions are aimed at removing the contamination before it becomes absorbed and lodged in the body. However, before decontamination can be started a careful survey must be carried out over the entire body with a suitable contamination monitor to determine the location of the contamination. In the case of partial contamination it is only necessary to decontaminate the affected areas. For instance, if a person has received contamination on the hands or face, these areas would be washed thoroughly with soap and water before being monitored again for residual contamination. Should contamination still be present the process is repeated until the affected areas are completely decontaminated.

In the case of whole-body contamination the first action, following removal of protective clothing, would normally be to wash the person's hair over a washhand basin. This removes contamination from the hair and prevents it being washed down into his or her mouth while undergoing whole-body decontamination under a shower using soap or more powerful cleansing agents. After the shower he or she is monitored carefully and may have to have further showers before decontamination is complete. In addition, when the person has been exposed to excessive airborne contamination, other measures, such as nose blowing and rinsing, may be required to remove the contamination completely.

If a minor wound is sustained in a contaminated area it should be allowed to bleed freely and be copiously washed with water to encourage the removal of the contamination. If the person cannot be immediately and completely decontaminated, or if the wound is serious, medical assistance should be obtained as quickly as possible. When the contamination has been swallowed, substances designed to prevent or reduce absorption from the gastrointestinal tract, e.g. antacids or ion exchange resins, may be administered promptly after the intake. If radionuclides of high toxicity, such as ^{239}Pu, are absorbed through a wound or inhaled in a soluble form, certain chemicals called 'chelating agents' (e.g. DTPA) may be administered to promote excretion. Unfortunately, these substances tend to be chemically toxic themselves. The absorption of certain radioisotopes can be 'blocked' by the prior ingestion of substantial amounts of a stable isotope of the same element. For example, the uptake of radioiodine to the thyroid can be greatly reduced by previous ingestion of a 200 mg tablet of potassium iodate. This has an important application in the event of a reactor accident.

9.9 CONTAMINATION MONITORING

9.9.1 Sensitivity

It has already been mentioned that quite small quantites of radioactivity which represent an insignificant external hazard can give rise to a significant internal hazard. This means that the radiation levels generated by an amount of radioactive contamination sufficient to cause an internal hazard are generally much lower than the levels which would cause an external radiation hazard. As a result contamination monitors require to be more sensitive, in general, than radiation survey meters.

To fulfil this requirement for increased sensitivity contamination meters are constructed from detectors which have their own built-in amplification system (Geiger-Müller or scintillation counters). The activity level is recorded as a counting rate (counts per second or per minute), and the monitor must be appropriately calibrated before the contamination level can be calculated (for example in MBq/m^2).

9.9.2 Direct surface contamination monitoring

This is the simplest and most convenient method of contamination monitoring, and is carried out to establish the presence of contamination on such surfaces as bench tops, clothing, skin and so on. Direct measurements allow the contamination level to be calculated in MBq/m^2, or they can be related to the derived limits of surface contamination. A typical contamination monitor consists of either a mains or battery operated ratemeter to which various types of detecting head can be connected.

Alpha contamination is detected by means of a zinc sulphide scintillator coupled to a photomultiplier tube. The zinc sulphide screen is covered with 'Melinex', a very thin plastic material, coated with aluminium to make it light tight. This covering has to be thin enough to allow alpha particles to penetrate through to the zinc sulphide screen. It is important when carrying out direct surface monitoring for alpha contamination to keep the probe as close to the surface as possible to record the true activity level.

Beta contamination monitors usually employ a Geiger-Müller tube in a suitable holder or probe. The probe has a shutter which should be opened for monitoring. If low energy beta contamination is suspected to be present, a thin end-window Geiger-Müller tube is used. Another type of beta detector uses plastic phosphors. These may be used in conjunction with a zinc sulphide screen to form a dual probe which permits simultaneous α- and β-monitoring. A single photomultiplier tube is used with a circuit to differentiate between the alpha and beta pulses.

Beta probes respond to gamma radiation which makes direct contamination monitoring difficult in areas of high gamma background. Under such circumstances indirect methods have to be used.

9.9.3   Smear surveys

Smear surveys are an indirect method of measuring surface contamination levels. They are used either to detect very low levels of contamination or to monitor for contamination in an area of high radiation background. A filter paper is wiped over a known surface area (usually $0 \cdot 1$ m^2), placed in a polythene envelope to avoid cross contamination and then taken to an area of low radiation background. Here it is counted in a detecting system of known efficiency. The contamination level can then be calculated from the formula:

$$\text{contamination level (Bq/m}^2) = C_c \times \frac{100}{E_c} \times \frac{1}{A} \times \frac{100}{E_F}$$

where

C_c = count rate, corrected for background, in counts per second

E_c = overall percentage efficiency of the counting system

A = area smeared in m^2

E_F = percentage of the contamination picked up by the paper.

The last quantity, E_F, is quite difficult to determine and is not very reproducible. It is dependent on various parameters, such as the physical and chemical nature of the contamination, the nature of the base surface and so on. In some circumstances E_F is taken as 100 per cent and in these cases it is the 'removable' contamination which is being determined. More usually a figure of 10 per cent is assumed.

A useful qualitative technique commonly used within active areas is to swab a large surface area using a damp paper towel and then to monitor the swab. This technique has the advantage that it also decontaminates the surface.

9.9.4 Air monitoring

Air monitoring is carried out in areas where airborne contamination may occur. There are basically three ways by which contamination can become airborne:

(a) by disturbance of surface contamination on the surfaces in the active area,
(b) by the drying out of liquid contamination,
(c) by dry, dusty operations, such as cutting, which cause the release of particulate activity.

Particulate airborne activity is measured by drawing a known volume of air through a filter paper. The filter paper is then counted in a low background area in precisely the same way as for smear survey papers. The level of airborne activity is calculated from the count rate on the filter paper by means of the formula:

$$\text{airborne contamination level (Bq/m}^3) = C_c \times \frac{100}{E_c} \times \frac{1}{V}$$

where

C_c = corrected count rate in counts per second

E_c = overall percentage efficiency of the counting system

V = volume of air sampled in m^3.

Gaseous activity is normally measured by drawing a certain volume of air through a filter paper into a sample chamber, which is then sealed. The filter paper removes particulate activity so that the activity in the sample chamber is due to radioactive gases only. The chamber is counted in a low background area and the gaseous activity level can be calculated.

9.9.5 Biological monitoring

Normally airborne and surface contamination monitoring are suffi-cient to ensure that insignificant amounts of contamination are entering personnel exposed to the contamination. In some circumstances, however, biological monitoring must be carried out. Examples of such circumstances are:

(a) the annual limit of intake of the radioisotope in question is very low (for example, plutonium),

(b) the isotope is difficult to detect by normal methods,

(c) an accident has occurred.

The type of monitoring used will depend on the type of radioisotope inside the body.

Gamma emitting isotopes can be measured in a whole-body counter, where the subject is placed in a low-background, shielded facility and the gamma emission detected by several large-volume sodium iodide (NaI) scintillation counters.

Alpha or beta emitting isotopes are measured by excretion monitoring, for example, faeces (ingestion of insoluble contamination), urine (soluble contamination). Breath monitoring is carried out to detect the inhalation of radium since the radium decays to its gaseous daughter product, radon, which is present in the exhaled breath.

Following a suspected inhalation of activity, monitoring of nasal swabs or nose-blows is a useful indicator of whether or not a significant intake has occurred.

SUMMARY OF CHAPTER

Internal radiation hazard: due to radioactive materials inside the body.

Routes of entry: inhalation, ingestion and direct entry through wounds in the skin.

Effective decay constant: $\lambda_{\text{eff}} = \lambda_r + \lambda_b$.

Annual limit of intake (ALI): intake of a radionuclide which would give a harm commitment to the organs it irradiates equal to that resulting from whole-body irradiation of 50 mSv (5 rem) in a year.

Derived air concentration (DAC): DAC = ALI/volume of air breathed in a year.

Derived limit (DL) for surface contamination: controls hazards from ingestion and direct irradiation.

Surface contamination monitoring: direct monitoring – zinc sulphide for alpha detection, Geiger-Müller tubes for beta detection. Indirect monitoring – using smear surveys.

$$\text{Surface contamination level (Bq/m}^2) = C_c \times \frac{100}{E_c} \times \frac{1}{A} \times \frac{100}{E_F}$$

Airborne contamination level: particulate contamination level determined by drawing a known volume of air through a filter paper:

$$\text{contamination level (Bq/m}^3) = C_c \times \frac{100}{E_c} \times \frac{1}{V}$$

gaseous activity measured by drawing a known volume of the atmosphere through a filter paper into a sample chamber and counting the activity.

Biological monitoring: (a) gamma emitters – whole-body counter;
(b) alpha or beta emitters – faeces, urine or breath monitoring.

REVISION QUESTIONS

1. Explain what is meant by the term annual limit of intake (ALI).

2. Using the values given in Table 9.2 and the conversion factors in Chapter 2, calculate the ALIs for sodium-22, iodine-131, caesium-137 and plutonium-239 in μCi.

3. If a worker is exposed to intakes of 5×10^5 Bq of iodine-131 (via inhalation) and 3×10^6 Bq of caesium-137 (via inhalation) in a year, what is the maximum external dose equivalent that he can receive in the year?

4. Calculate the maximum intake of cobalt-60 dioxide which a worker may receive via inhalation in a year if his dose equivalent from external exposure is 40 mSv (4 rem). The inhalation ALI for cobalt-60 dioxide is 10^6 Bq.

5. Calculate the derived air concentration for americium-241. The ALI (inhalation) for ^{241}Am is 2×10^2 Bq.

6. Discuss the problem of direct surface contamination monitoring.

7. Calculate the surface contamination level from the following data:

Uncorrected count rate on smear paper 3840 counts per minute
Background count rate   240 counts per minute
Efficiency of counting system   15%
Area of surface smeared $0 \cdot 1 \text{ m}^2$
Pick-up efficiency of smear   10%

8. Calculate the level of airborne particulate activity from the following data:

Uncorrected count rate on filter paper   29832 counts in 4 minutes
Background count rate   1032 counts in 4 minutes
Efficiency of counting system   12%
Flow rate through particulate sampler $0 \cdot 03 \text{ m}^3$ per minute
Time for which sampler was run   2 minutes

9. Discuss the problem of dealing with a contaminated person who is also injured.

10. What is meant by the 'radiotoxicity' of an isotope? Discuss how this affects the classification of laboratories which handle radioactive material.

10 —————————————————

Nuclear reactor
health physics

10.1 INTRODUCTION

The discovery of fission in 1938 provided the basis of a new source of energy potentially greater than the entire world reserves of fossil fuels. The development of nuclear energy has been quite rapid. The first fission reactor was operated in a converted squash court in Chicago in 1942 by Enrico Fermi. The rate of subsequent development has been such that nuclear energy now supplies a significant fraction of the power requirements of most of the advanced countries of the world. The advantages of the nuclear reactor as a source of power are offset to some extent by a number of special problems. These include:

(a) the protection of the operator and maintainer,
(b) the safe treatment and disposal or storage of the radioactivity produced, and
(c) the need to achieve an acceptably low risk of injury to the public due to potential large releases of radioactivity in the event of a reactor accident.

None of these problems is insurmountable and it is generally agreed that the nuclear energy industry has an excellent safety record.

In order to understand the hazards associated with reactors, a basic knowledge of nuclear fission and reactor technology is necessary.

In this chapter, after a basic discussion of the process of nuclear fission, and of reactor technology, the radiological hazards involved in the operation of nuclear reactors are outlined.

10.2 FISSION

10.2.1 The fission process

Fission is the splitting of a nucleus into two approximately equal parts, known as fission fragments. Certain types of heavy nuclei, notably uranium and thorium, are found to undergo spontaneous fission at a rather low rate. Others can be made to fission by the addition of energy, for example by bombardment with neutrons. Materials which can be made to fission in this way are said to be fissile. The process of fission results in the release of energy, mainly in the form of kinetic energy of the fission fragments. This is rapidly converted into thermal energy and raises the temperature of the fuel material. Of naturally occurring materials, only the isotope uranium-235 (^{235}U) is fissile to a significant extent. This constitutes only 0·7% of natural uranium, the remaining 99·3% being uranium-238.

A fundamentally important feature of fission is that the fission fragments are so unstable that they give off neutrons, usually between 2 and 4 neutrons per fission (Fig. 10.1). Most of the neutrons are emitted almost instantaneously, and are called prompt neutrons, but some are released some seconds or even minutes after fission – these are delayed neutrons.

The emission of neutrons by the fission process has a number of consequences. First, it makes possible a chain reaction; second, neutron capture reactions in the uranium fuel result in the production of transuranic elements, including plutonium (see Section 10.2.4); and third, neutron captures in the structural and other materials of nuclear reactor plants result in these materials becoming activated (see Section 10.2.5).

10.2.2   Chain reactions and criticality

The neutrons emitted following fission may themselves cause further fission with the emission of more neutrons, so making possible a chain reaction. In practice some of the neutrons escape from the system and others are captured in non-fission reactions. In a reactor, wastage of neutrons by capture reactions is minimized by taking care to exclude materials of high capture cross section from the core. Some loss is

Fig. 10.1   Fission

inevitable since even the fuel material can capture neutrons without fission. By increasing the size of the core the fraction of neutrons escaping can be reduced to a sufficiently low level to permit a chain reaction. If a core is very small, neutrons can easily escape but in larger cores they would require to travel further and are, therefore, more likely to cause fission. Another method of reducing leakage is to put a reflector of some light material around the core to reflect escaping neutrons back into the core.

Consider an assembly of fissile material in which, on average, 2·3 neutrons are produced by each fission. If, on average, 1·3 neutrons are lost by leakage or capture, one neutron is available to cause a further fission. This results in a self-sustaining chain reaction in which the fission rate is constant. Such a system is said to be *critical*. An average loss of 1·31 neutrons for every fission means that only $2·3 - 1·31 = 0·99$ neutrons are available for fission. The system is then subcritical and the fission rate will decrease. Conversely, for a loss of 1·29 neutrons per fission, 1·01 neutrons remain. This results in a supercritical system in which the fission rate will increase.

Thus for any given type of reactor there is a minimum size of core below which the system cannot go critical. Alternatively, a certain minimum mass of fuel is required for a chain reaction to be possible, this is the minimum *critical mass*. For example, a nuclear weapon can, in principle, be made from two pieces of fissile material each of which is slightly more than one half of the critical mass. The weapon is detonated by bringing the two pieces rapidly together to give a supercritical mass. This results in the fission rate, and hence the rate of energy release, rising very rapidly to enormous values. Fortunately this is rather difficult to achieve in practice!

The possibility that fissile material could 'go critical' if assembled in a sufficient quantity has very important implications in the design and operation of plants in which it is processed, handled or stored. It is obviously very important to ensure that criticality does not occur and this is usually achieved by careful design of the facilities supplemented, in some cases, by operational procedures. This aspect is discussed further in Chapter 16.

10.2.3 Fission products

In fission, the splitting of the atoms of a fissile material can occur in many different ways. The most likely division of the heavy atom is into fragments of mass about 97 and 135. For example,

$$^{235}_{92}U + ^{1}_{0}n \rightarrow ^{135}_{52}Te + ^{97}_{40}Zr + 4^{1}_{0}n$$

The distribution of mass number of fission products is of the form shown in Fig. 10.2. The fission process produces about 300 different nuclides, most of which are neutron rich and decay by a series of beta emissions through a decay chain of radioactive nuclides. In the example of fission shown above, the tellurium-135 and zirconium-97 both undergo a series of beta decays until a stable nuclide is produced:

$$^{135}_{52}Te \rightarrow ^{135}_{53}I \rightarrow ^{135}_{54}Xe \rightarrow ^{135}_{55}Cs \rightarrow ^{135}_{56}Ba$$

and

$$^{97}_{40}Zr \rightarrow ^{97}_{41}Nb \rightarrow ^{97}_{42}Mo$$

It will be seen that the utilization of nuclear fission to produce energy results in the formation, within the fuel, of hundreds of different types of radioactive fission products with half-lives varying from a fraction of a second to very many years. The inventory of fission products in the fuel builds up over the period of irradiation.

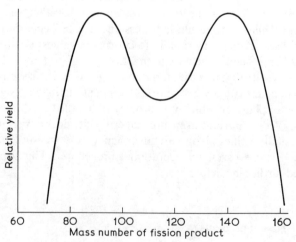

Fig. 10.2   Relative fission product yield

Fig. 10.3 Approximate fission product beta inventory in fuel, irradiated for 2000 days at 1 MW

Figure 10.3 shows the fission product inventory, in becquerels and curies, and its decay from 1 to 1000 days in fuel which has produced a uniform 1 megawatt of heat over a 2000 day irradiation.

The many different nuclides present as fission products have a very wide range of physical and chemical properties and so the radiotoxicity of the different nuclides varies enormously. In order to prevent the escape of the fission products, the fuel is contained within another material of suitable properties, known as the *cladding*.

10.2.4 Transuranic elements

As previously noted, neutrons can be captured by fuel atoms in non-fission reactions. Successive capture reactions and beta decays result in the generation in the fuel material of transuranic elements, that is elements higher in the periodic table than uranium. Of particular importance is neutron capture in ^{238}U yielding ^{239}U, which quickly undergoes beta decay to neptunium-239 (^{239}Np), and then to plutonium-239 (^{239}Pu):

$$^{238}_{92}U(n, \gamma)^{239}_{92}U \xrightarrow{\beta} {}^{239}_{93}Np \xrightarrow{\beta} {}^{239}_{94}Pu$$

This is a process of great importance since it provides a route whereby a large proportion of ^{238}U, which is not fissile to a significant extent, can be converted into ^{239}Pu which is fissile. However, ^{239}Pu has the disadvantage of being a long-lived ($T_{1/2} = 24\ 300$ years) α-emitter and, unlike uranium, it is of very high radiotoxicity.

Further capture and beta decay processes lead also to the production of americium (Am) and curium (Cm) nuclides.

10.2.5 Activation products

Neutron capture in structural materials and in the coolant of a reactor results in the generation of many species of radioactivation products. In steel components, the stable iron isotope ^{58}Fe undergoes neutron capture to become ^{59}Fe, which decays by β-emission to ^{59}Co with a half-life of 45 days.

In many cases, because of a particular combination of nuclear properties, activation of trace elements in a material is important. For example, all types of steel contain cobalt, normally at a concentration of only a few hundred parts per million. The stable isotope ^{59}Co is activated by the reaction

$$^{59}\text{Co}\,(n, \gamma)^{60}\text{Co}$$

Cobalt-60 has a relatively long half-life of 5·22 years and, in addition to a low energy beta particle, emits two energetic gamma rays per disintegration.

In both water-cooled and gas-cooled reactors, an important reaction on the oxygen component of the coolant is

$$^{16}\text{O}(n, p)^{16}\text{N}$$

Although nitrogen-16 has a very short half-life of 7·2 s, it has a great influence on the shielding requirements.

10.3 REACTOR SYSTEMS

10.3.1 General features

The great majority of commercial nuclear power reactors currently operating or under construction, worldwide, are light water reactors (LWR), which were originally developed in the USA. These are of two types, pressurized water reactors (PWR) and boiling water reactors (BWR). In a PWR a closed coolant system transfers heat from the core to a heat exchanger, which raises steam in a secondary circuit to drive a turbogenerator (see Fig. 10.4). In the BWR, the water passing through the core is allowed to boil and the steam is passed directly to the turbines. Another type of water-cooled reactor is the Canadian CANDU system which uses heavy water.

Fig. 10.4 Primary system of a PWR

In the United Kingdom, a different line of development has been pursued and has resulted in two generations of gas-cooled reactors. In the earlier type, known as the Magnox reactor, the core consists of a very large graphite structure, penetrated by channels containing natural uranium fuel rods clad in a magnesium alloy and cooled by carbon dioxide. The carbon dioxide circulates through a heat exchanger, where steam is raised as in a PWR. The second generation UK plants are known as advanced gas-cooled reactors (AGR). In order to achieve a higher operating temperature, stainless steel fuel cladding is used and this necessitates the use of enriched uranium, as discussed in the following section. Typical Magnox and PWR plants are illustrated in Figs 10.5 and 10.6.

Reactor   Heat exchanger   Turbine   Generator

Condenser

Steam

Water

Gas blower

Hot gas duct

Cool gas duct

Charge tube

Control rod

Fuel element

Graphite moderator

Concrete shield

Fig. 10.5 Schematic illustration of a Magnox power reactor

Fig. 10.6 Schematic illustration of a pressurized water reactor

Another system being developed, particularly in France and the United Kingdom, is the sodium-cooled fast breeder reactor, sometimes known as the liquid-metal fast breeder reactor (LMFBR). The early fast reactors will utilize plutonium fuel extracted from spent thermal reactor fuel but, later, will breed enough plutonium by neutron capture in ^{238}U in and around the core to meet their own fuel requirements.

The central feature of any nuclear reactor is the core, which contains the fissile material, a means of controlling the fission rate and a moderator in the case of thermal reactors but not in the case of fast reactors. The other main features are a cooling system to remove the heat generated by fission and a radiation shield, often called the biological shield. These various aspects are described below.

10.3.2   The core and control system

The reactor core contains fuel elements which consist of fissile material in a fuel can or covered by a cladding material which prevents the escape of the fission products. The fuel is usually uranium although plutonium and thorium are sometimes used. Naturally occurring uranium contains 0·7 per cent of the isotope ^{235}U and 99·3 per cent of ^{238}U. Magnox and CANDU systems are able to operate using natural uranium, but in other systems it is necessary to use uranium containing a higher percentage of ^{235}U and this is achieved by means of an enrichment process.

The moderator serves to slow down the fast neutrons produced by fission to the thermal energy range in which they are most likely to cause further fission. In water reactor systems, the water serves both as moderator and coolant, whilst in the gas-cooled systems the moderator is graphite.

The control rods are made from materials of high neutron absorption cross section such as boron or cadmium. Withdrawal of the control rods to a certain minimum position takes the reactor critical. Further withdrawal makes the reactor supercritical and the fission rate increases. When the required power is reached, insertion of the rods to the critical position causes the power to become constant. The reactor power can be reduced by insertion of the rods to a subcritical position. The control system is based on a number of neutron detectors around the core. If the neutron flux exceeds a pre-set value or if it is increasing too rapidly the reactor is automatically shut down by the rapid insertion of the control rods (or sometimes special shut-down rods). Various other monitoring systems provide data which can shut down the reactor if something goes wrong, for example, high gamma radiation levels outside the biological shield, excessive core or coolant temperature, or loss of coolant flow.

10.3.3 The cooling system

Fission causes energy release within the fuel and consequent tempera-ture rise of the fuel and cladding material. The heat is removed by the coolant and steam is raised, directly or indirectly, to drive turbines. Various subsidiary systems are provided for coolant purification, sampling and so on.

The integrity of the cooling system must be very high. Most coolants become radioactive to some extent and even small leaks may pose radioactive contamination problems. Large leaks could limit the heat removal from the core and affect the safety of the reactor. It is important to realize that in a power reactor the decay of the fission products generates a substantial amount of heat and so cooling must be guaranteed even when the reactor is shut down.

10.3.4 The biological shield

The purpose of the biological shield is to attenuate the neutron and gamma radiation from the reactor core and cooling system so that the operators and maintenance staff do not receive excessive doses. The most commonly used materials are lead, concrete, iron, water and polythene. In the case of power reactors the factor of reduction required may be of the order of 10^8 to 10^9. Weaknesses can arise in the shield due to streaming paths of radiation down coolant ducts and other penetrations. The shield is often constructed in two parts – the primary shield around the reactor core and the secondary shield around the cooling system. This arrangement permits access to the coolant system (circulators, heat exchangers, etc.) when the reactor is shut down since the gamma radiation from the decay of fission products is attenuated by the primary shield.

10.4. REFUELLING REACTORS

In a nuclear reactor, one kilogram of natural uranium (about one cubic inch) can produce as much energy as 10 tonnes of coal burned in a conventional power station. This is in spite of the fact that only 0·3 per cent of the uranium atoms are burned up, which represents about 40 per cent of the fissile ^{235}U atoms. The percentage of uranium which can be burned up is limited, since a stage would be reached when insufficient fissile material is present to sustain a chain reaction. In addition, pressures generated by the fission products in the fuel cause swelling and distortion of the fuel elements. When a fuel element reaches the required burn-up, it must be removed from the reactor (remember that it is intensely radioactive), and replaced by a fresh element.

In the large gas-cooled reactors, refuelling of the reactor is performed with the reactor operating. The operation is performed by a remotely controlled charge/discharge machine on top of the reactor which removes the shielding plug, seals itself on to the pressure vessel and removes the fuel channel cap. The spent fuel elements are raised from the channel into the machine, which is heavily shielded, and replaced by fresh elements. When the machine contains its full quota of elements it is moved to a position where the elements can be released through a shielded chute into a large tank of water, known as a *cooling pond*, which provides shielding and cooling for the spent elements. The machine is, of course, designed to make these operations completely safe. It must, therefore, provide shielding from the reactor core above the unplugged hole, provide shielding from the spent fuel elements inside the machine, prevent leakage of gas or contamination and also provide cooling for the fuel elements (the heat generated by decay of the fission products in an element when it is removed from the reactor is sufficient to cause it to overheat unless external cooling is provided).

In LWR, on-load refuelling is not possible for a number of reasons and normally the reactor is shut down annually and about one third of the core is replaced. This involves flooding a canal above the reactor pressure vessel, removing the pressure vessel lid and lifting the fuel elements out one at a time into the water filled cavity. They are then transferred down the canal into a storage pond.

10.5   RADIATION HAZARDS FROM REACTORS

10.5.1   General

In general reactors are a lesser radiation hazard when they are operating than when they are shut down. The shielding is designed to give acceptable radiation levels at positions where men are required to work and thorough surveys are made during the commissioning of the plant and subsequently at regular intervals. Systems are provided which permit safe means of sampling coolants and other radioactive effluents. During shutdown periods, a great variety of non-routine jobs may be undertaken, some of them on highly radioactive systems. It is during such periods that exposure of personnel to radiation and radioactive contamination must be carefully controlled.

A serious fault or mal-operation could cause considerable damage to the plant and give rise to dangerously high levels of radiation or radioactive contamination. If the hazard is confined to the reactor site it is often called a *site emergency* but if it extends off-site it may become a *public emergency*. Accidents of this type are covered in Chapter 16.

10.5.2 Sources of radiation

The main sources of radiation from a reactor at power are the core and the coolant. The radiation from the core includes fission neutrons, fission gamma rays, fission product decay gamma rays, neutron capture gamma rays and activation product decay gamma rays. The latter two arise predominantly in the core structure and shield. The radiation from the coolant is mainly gamma rays due to activation of the coolant, activation of impurities and fission product contamination of the coolant. The sources are illustrated in Fig. 10.7.

Radiation from the core

The neutrons produced in a reactor as a result of the fission process are fast neutrons in the range 0·1 to 15 MeV, average energy about 2 MeV. Those emerging from the surface of the biological shield have undergone varying degrees of moderation and so neutrons of all energies from thermal to fast may be present.

Fission gamma rays are those emitted immediately after the fission fragments and vary in energy from 0·25 to about 7 MeV. The gamma radiation resulting from the decay of fission products in the fuel elements is small compared to the fission gamma radiation but whereas the latter ceases on shutdown of the reactor the fission products continue to emit radiation for many years after the fuel has been withdrawn from the reactor.

Neutron capture in the structural materials of the reactor and in the shield results in the emission of capture gamma rays and also makes

Fig. 10.7   Sources of radiation in a nuclear reactor system

these materials radioactive. The radiation from the decay of the radioactivity, as in the case of fission products, continues to be emitted when the reactor is shutdown. Neutron scattering also leads to gamma ray emission but does not, in general, induce radioactivity.

Radiation from the coolant

As noted previously, in both water-cooled and CO_2-cooled reactors, the reaction

$$^{16}O(n, p)^{16}N$$

is important because nitrogen-16 decays with a half-life of 7·2 s and emits very penetrating (6·1 MeV) gamma rays. This means that the coolant circuit of the reactor is a significant source of gamma radiation and, in some cases, must be shielded.

Air was used as a coolant in a number of reactors of early design and is used on Magnox reactors to cool the biological shield. Air is also present as an impurity in the coolants of both gas- and water-cooled reactors. The argon which is present to the extent of 1·3 per cent in air is activated by the reaction:

$$^{40}Ar(n, \gamma)^{41}Ar$$

Argon-41 decays with a half-life of 1·8 h and emits gamma radiation of energy 1·29 MeV.

In LMFBR, which use sodium as a coolant, the important activation products are sodium-22 (^{22}Na) and sodium-24 (^{24}Na).

Although reactor coolants are very pure by normal standards, there are always some impurities. When subject to the very high neutron flux in the reactor core, these impurities become radioactive to an appreciable extent. Impurities also arise as a result of corrosion or erosion from the core and the coolant system. For example, in Magnox and advanced gas-cooled reactors, graphite dust (which itself contains impurities) collects around the cooling system. In water-cooled reactors, iron, nickel, cobalt and manganese are common impurities due to corrosion of the coolant system. Deposition of this corrosion material in the core causes the build up of a film of corrosion products on fuel surfaces. As a result of irradiation by neutrons, these films become highly radioactive. The continuous release of material from the core and subsequent deposition in the out-of-core regions causes a build up of radioactivity in the coolant system.

Reactor coolants usually contain readily detectable levels of fission product contamination. This can occur in three ways:

(*a*) Uranium contamination on the fuel element surface.
(*b*) Uranium impurity in the fuel cladding material.
(*c*) Release from a damaged fuel element.

The coolant is being continually cleaned up by a treatment system and so the long-lived fission products do not build up appreciably. The predominant fission product activities are usually krypton-88 and xenon-138 which are inert gases, their particulate daughter products rubidium-88 and caesium-138, and the isotopes of iodine, ^{131}I, ^{133}I and ^{135}I. A seriously damaged fuel element could lead to a considerable fission product activity being spread around the cooling system. Most gas-cooled reactors are fitted with a system known as 'burst can detection' which continuously and automatically 'sniffs' each channel in turn and gives warning to the operators if an increase in the fission product activity should occur. The channel containing the damaged fuel can is then unloaded and the faulty element is replaced.

10.5.3   Sources of radioactive contamination

Beta emitters

Almost all of the radioactive nuclides mentioned in the preceding paragraphs decay by beta emission. Beta radiation is so easily absorbed that the shielding designer does not even need to consider it, concentrating on the associated gamma emission. If radioactive contamination occurs, due to a leak of radioactivity from the reactor system, the beta radiation is often of prime importance.

A radionuclide of considerable importance which is produced in all reactor systems is tritium (^{3}H). This nuclide has a half-life of 12·3 years and decays by low energy beta emission only. It is produced by fission, by a (n, γ) reaction on deuterium (hydrogen-2), and by various reactions on lithium and boron. In reactors cooled or moderated by heavy water, large amounts of tritium build up. In light water systems, because of the much lower concentration of deuterium and the frequency of water change (heavy water is much too expensive to change), build up of tritium is usually less significant. Lithium and boron are present in most systems either as additives, neutron absorbers or impurities and can contribute to tritium production by a number of reactions including

$$^6\text{Li(n, }\alpha)^3\text{H} \quad \text{and} \quad ^{10}\text{B(n, }2\alpha)^3\text{H}$$

Coolant leaks

Contamination can, of course, occur because of a coolant leak. In pressurized water systems the leak may be direct to atmosphere or via a heat exchanger into the secondary system in which case radio-activity, mainly the gaseous activities ^{88}Kr, ^{138}Xe and ^{41}A, would be carried over with the steam into the turbines and then to atmosphere via the condenser air ejector. The fission product gases ^{88}Kr and ^{138}Xe decay to their particulate daughters ^{88}Rb and ^{138}Cs. A leak in the heat exchanger of a gas-cooled reactor would normally cause steam to leak into the primary system because of the higher secondary pressure.

Containment

The core of a reactor at power contains about 0·2 TBq of fission products per watt of thermal power. Thus a reactor operating at, say,

1000 megawatts (MW) contains about 2×10^8 TBq (~ 5000 MCi) of fission products. This vast inventory of radioactivity is contained within the fuel can or cladding which provides the first level of containment. The second level of containment is the boundary of the primary systems, that is, the pressure vessel and the coolant system. This boundary also contains the radioactivity of the coolant which, in a 1000 MW reactor, may amount to some hundreds of TBq. Provided that the primary and secondary containments remain intact there is little risk of serious contamination. In practice some contamination does occur during the operation of most reactors. It is usually necessary to sample the coolant periodically and there is often some radioactive effluent. A contamination hazard could arise in both cases but the risk is minimized by good design of facilities.

In its most common usage, the term 'reactor containment' refers to the structure within which the whole system is housed. An example of such a containment is the familiar sphere of the Dounreay fast reactor. The function of this containment is to protect the general public by limiting the release of fission products in the event of a serious reactor accident. (See Chapter 16.)

10.5.4 The shutdown reactor

Maintenance

A reactor represents a large capital investment and shutdown periods, whether scheduled or not, are costly. Reactor systems are comparatively simple, well engineered and normally very reliable. The majority of maintenance, corrective and preventive, is on ancillary or secondary equipment. A major overhaul may include decontamination and refitting of coolant circulators, control rod mechanisms, inspection of heat exchangers and various other jobs on radioactive systems. At such times, the need to keep to a tight schedule can lead to a general reduction in standards of safety, both radiation and conventional, because it may slow down the work. In order to prevent this state of affairs all major work is planned in consultation with interested parties and sometimes it is desirable to rehearse particularly difficult jobs. Personnel should receive instruction in general safety matters and be familiar with the particular hazards associated with their own work.

External radiation

When the reactor is shut down, the primary shield gives adequate protection from the fission products in the core. The radiation hazard to personnel working on the primary system is due to radioactivity within the system. The dose rate in the vicinity of the primary system tends to decay rapidly in the first 24 hours after shut down, mainly due to decay of coolant activities or their clean-up by the treatment system. Thereafter, the levels do not change significantly from day to day. The half-lives of most of the radioactive corrosion products are in the range 1 month to about 5 years. The dose rates vary considerably from reactor to reactor but in systems with corrosion problems, levels of 10 to 100 mSv/h (1 to 10 rem/h) are frequently encountered on certain components. If the dose rate is excessive it is sometimes possible to provide additional shielding on 'hot-spots'. An alternative is to decontaminate the component but this would only be done during major shutdown periods.

Careful control is required of personnel working in areas of high dose rate. This often takes the form of a manned control point at the entrance to the area. Personnel entering the area are given a 'working time' based on a radiation survey of the area. In addition to their normal film badge they are required to wear some form of direct reading dosemeter such as a quartz fibre electroscope (QFE). The times of entry and exit and the QFE reading are logged.

Contamination

When maintenance involves breach of the primary coolant system, contamination is likely to occur. When first breaching part of the system it is obviously essential to have depressurized. It is good practice for the personnel involved to wear full-face masks when first breaking into part of the system.

The standards of protective clothing required are evaluated from experience on a particular plant. There are often other factors such as temperature, humidity and possible presence of toxic gases, all of which affect the choice of protective clothing and equipment. Personnel cannot be expected to wear impervious clothing, such as PVC suits, in temperatures of 40–50°C, unless the suits are fully ventilated. This in turn causes difficulties in confined spaces because of the required air-lines. Efficient changing and monitoring facilities are essential. If significant levels of contamination are present, it is advisable to have an attendant to assist in the removal of contaminated clothing. This can often be the same person who logs the entry of personnel.

Frequent monitoring of levels of contamination both inside and outside the area is undertaken.

10.6 RESEARCH REACTORS

Research reactors, of which there are many different types, present special health physics problems. The reactors have a wide range of application including fundamental research, materials testing and the commercial production of radioactive sources. The main radiological problems arise not from the reactor, but usually from the experimental equipment. For example, there are often holes through the shielding in order to obtain high intensity beams of neutrons outside the reactor for various purposes. Rigs containing experimental equipment or materials become highly radioactive. In these and other cases special precautions are necessary to protect both operators and users of the reactor.

10.7 FUEL STORAGE PONDS

10.7.1 Introduction

After removal from a reactor, it is usual to allow the fuel to decay for some months in a fuel storage or cooling pond situated close to the reactor. This procedure eases some of the problems involved in moving the highly radioactive fuel from the reactor site to the fuel reprocessing plant. In addition to the routine operational problems, ponds pose two special hazards – criticality and loss of shielding accidents.

10.7.2 Criticality

Ponds often contain enough fissile material to 'go critical', that is to initiate a fission chain reaction. In general, if only natural uranium fuel is in the pond criticality is not possible even under the worst possible conditions. If enriched uranium or other fissile material such as plutonium is present, criticality could conceivably occur. The hazard is controlled by storage in safe configurations, that is with adequate spacing between fuel elements or by limiting the number of elements which may be out of their storage position at any time.

10.7.3   Loss of shielding

A typical storage pond may often contain thousands of TBq of fission product activity and a single element may contain a few hundred TBq. Such an element unshielded would give a gamma dose rate in excess of 1 Sv/h (100 rem/h) at a distance of 3 m. Obviously this situation must not be allowed to occur. Loss of shielding can occur by loss of pond water or by raising fuel too close to the surface of the water. Loss of pond water is more likely to be due to accidental pumping out than to a leak caused by serious structural damage. The possibility of pumping out can be minimized by a good design of the water system and by administrative control of the system (e.g. locks on vital valves).

It is difficult to safeguard completely against a fuel handling accident without defeating the advantages of water shielding, such as visibility and flexibility. Good design of lifting tackle can make it impossible to raise fuel too high but only if the correct tackle is used. A little ingenuity can defeat the best systems. Considerable reliance is placed on installed gamma monitors to give warning of potentially dangerous situations.

10.7.4   Operational aspects

The fission product activity of the fuel is contained within the fuel cladding. A breach of the cladding would lead to contamination of the water by fission products to some extent depending on the form of the fuel. As a precaution against the release of gaseous fission products, which would be rapidly released from the water surface, many ponds have an air extract system 'sweeping' the water surface. Some degree of water contamination also occurs due to the release of radioactive corrosion products from the surface of the fuel cladding. To prevent build up of activity a water treatment system is provided. A water cooler is sometimes necessary, depending on the amount of fuel stored.

Cutting operations in a pond can lead to serious contamination problems due to swarf, particularly if fuel is cut. The maintenance of the cutting equipment may be hindered by high contamination and radiation levels.

10.7.5 Pond instrumentation

There are three main types of instrument used to give warning of hazardous conditions:

(a) Installed gamma monitors are an essential feature of any fuel storage pond. At least three instruments are necessary and they should be situated so that they cannot be shielded from the pond surface. A local alarm such as a bell or buzzer should sound if any of the instruments reach some quite low level, say 0.1 mSv/h (10 m rem/h). This may warn the operators that a potentially dangerous situation is developing. At some quite high radiation level, say 0.1 Sv/h (10 rem/h) *at the operators position*, an evacuation alarm should sound. This alarm should operate on a two out of three or two out of four basis to avoid spurious evacuations.

(b) Pond water counter: this is a simple device which continuously monitors the water for beta activity and gives early warning of damaged fuel. A shielded liquid flow Geiger-Müller tube connected to a ratemeter is normally used.

(c) Air monitor: a continuously operating airborne particle monitor in the general area or in the air extract gives warning of high airborne contamination arising from pond operations or maintenance of pond equipment.

10.8   FUEL REPROCESSING

After a period of pond storage at the reactor site to allow some decay of the levels of radioactivity and decay heat, irradiated fuel is transported to a fuel reprocessing plant. Here, after a further period of pond storage, the fuel is dissolved in nitric acid and the solution is chemically separated into three main streams, unused uranium, plutonium and a highly active waste stream containing almost all of the fission products and higher actinides such as americium and curium.

A reprocessing plant poses much greater radiological problems than a reactor, because of the nature of the processes involved. Essentially, a reprocessing plant is a complex chemical facility, processing intensely radioactive solutions which must be shielded from operating areas by a metre or more of concrete. The plant requires maintenance and although facilities are provided to enable much of it to be done remotely, some contact maintenance is inevitable. Although the part of the plant requiring maintenance will have been decontaminated, this is never completely effective. In addition relatively small leaks of the radioactive solutions into the shielded cells of the plant cause severe radiation and contamination problems, and decontamination of cell and equipment surfaces is very difficult and time consuming.

Of the product streams, the uranium stream does not present any significant radiological problem but the plutonium stream demands extremely high standards of containment to avoid a severe contamination, particularly an inhalation, hazard.

In addition to the highly active waste stream, which is routed into special storage tanks, there are a number of subsidiary radioactive waste streams, the treatments of which depend on local conditions. The question of radioactive waste is covered in the following chapter.

Another important factor in the design and operation of fuel cycle facilities is the possibility of a criticality accident. Such an accident could result not only in high neutron and gamma radiation from the event, but also, because of the energy generated, in dispersal of the process material.

SUMMARY OF CHAPTER

Fission: splitting of nucleus into two fission fragments which decay to fission products. Neutron chain reaction – critical mass.

Reactor system: core consists of fissile material, control rods and moderator; other features are a cooling system and a biological shield.

Refuelling: on- or off-load.

Sources of radiation when operating: fission neutrons and γ-rays, fission product decay γ-rays, neutron capture γ-rays and activation product decay γ-rays.

Shutdown sources: fission and activation product decay γ-rays.

Contamination may occur due to coolant leak or maintenance operations.

Containment of fission products.

Fuel storage ponds: two special hazards – loss of shielding and criticality.
Fuel reprocessing plant: chemical separation of fuel into uranium, plutonium and waste streams. Severe radiological problems.

REVISION QUESTIONS

1. Describe the process of nuclear fission and explain the circumstances under which a fission chain reaction may be achieved.

2. Draw a simple sketch of a typical nuclear reactor and label the major features of the system.

3. Using a simple diagram, illustrate the sources of radiation from a nuclear reactor system, indicating which are important when:

   (*a*) operating, and
   (*b*) shutdown.

4. What is meant by *containment* in the context of nuclear reactors?

5. Discuss the main hazards presented by fuel storage ponds and how they are controlled.

6. What is a fuel reprocessing plant and what are the main radiological hazards encountered?

11
Radioactive waste

11.1 INTRODUCTION

Prior to the discovery of nuclear fission and its utilization as a source of energy, the disposal of radioactive waste did not present a significant problem. It has been estimated that the total quantity of radioactivity in use in research and medicine in 1938 was less than 30 TBq, corresponding to about 1 kg of radium derived from natural sources. Today, a single large power generating reactor may contain in excess of 10^8 TBq of fission products and there are several hundreds of power reactors in the world. With the increasing emphasis on protection of the environment, the management of these wastes is becoming an important factor in both the economics and the public acceptability of nuclear power.

On a much smaller scale, radioactive wastes arise in hospitals, factories, research and teaching institutions as a result of the many applications of radiation. In such cases, the complex treatment plant used at nuclear power stations would be prohibitively expensive and so disposal may be by the normal refuse collection or sewage systems. Clearly, the consequences of such practices must be understood and strict control must be exercised.

Radioactivity cannot be destroyed. It will decay eventually, but in view of the very long half-life of many radionuclides, it is not always practicable to await the decay of radioactive material. There are three general approaches to the management of radioactive wastes:

(a) *release* and *dispersal* into the environment,
(b) *storage*, and
(c) *disposal*.

Of course, release of radioactivity to the environment might reasonably be thought to constitute disposal. However, it is useful to distinguish between deliberate dispersal and methods of disposal involving irretrievable placement of wastes so that they are isolated, at least temporarily, from the environment.

Storage of radioactive waste is a useful procedure when dealing with nuclides of relatively short half-life (e.g. up to a few months). Storage or hold-up of the waste for a period of up to a few years may reduce the activity to a sufficiently low level to permit release to the environment or, in the case of solid waste, to facilitate disposal.

The approach selected in a given situation depends on many factors such as the quantity and type of radioactivity, its physical and chemical forms and the geographical location. In this chapter, after a discussion of the consequences of releases of radioactivity, the application of these alternative approaches to the management of liquid, gaseous and solid wastes is discussed.

11.2 CONSEQUENCES OF RELEASES OF RADIOACTIVITY

Any release of radioactivity is a potential source of radiation exposure to the population at large. The radiation exposure can occur via many different *exposure pathways*. Consider, for example, a release of activity from a chimney stack. This would be dispersed by air movements and could result in radiation exposure of the population in a number of ways;

(a) direct external β- or γ-radiation from the plume,
(b) inhalation of radioactivity resulting in internal dose,
(c) direct external β- or γ-radiation from deposition (fall-out) of radioactivity,
(d) consumption of foodstuffs (e.g. vegetables) contaminated by deposition, and
(e) consumption of meat or milk from animals which have grazed on contaminated ground.

Similarly, radioactivity discharged into a river, lake or sea could result in a number of exposure pathways:

(a) contamination of drinking water supplies,
(b) external dose to swimmers or to people using contaminated beaches, and
(c) consumption of contaminated fish, shellfish or plants.

Some of these exposure pathways result from complex routes known as *food chains*. For example, activity discharged to the sea may be taken up by plankton which are eaten by fish, which are in turn eaten by man. An example of a simple marine food chain is illustrated in Fig.11.1.

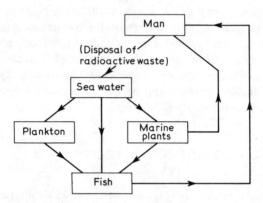

Fig. 11.1   A simple marine food chain

In a given situation it is often found that one particular pathway is much more important than any other. That is, it results in much more radiation dose, sometimes to quite a small group of people. This is called the *critical exposure pathway* and the group of people which receives the highest dose from this pathway is known as the *critical group*. The importance of the critical exposure pathway is that the quantity of radioactivity discharged must be controlled so as to limit the dose to the critical group. The critical group in any particular case depends on the mode of release, the nuclides involved, local ecology (e.g. forms of marine life, etc.) and local habits.

The fuel reprocessing plant at Sellafield, in the UK, provides a classic example of the application of this concept. In the plant, irradiated fuel from reactors is processed chemically to recover uranium and plutonium. The other elements including the fission products and higher actinides, which are intensely radioactive, are transferred to special storage facilities (see Section 11.5 below). However, a small proportion of the activity appears in subsidiary waste streams, which after a period of storage to permit some decay are discharged into the Irish Sea via a 2 km pipeline.

Before any activity was released from the plant a detailed study was made in order to estimate how much activity could safely be discharged. This involved investigations of the local tidal currents and the dispersion which would occur, and of the local ecology. This preoperational study suggested that there would be three main exposure pathways and these involved deposition of activity on sand and silt, uptake in fish, and uptake of activity in a variety of edible seaweed. It was anticipated that the latter pathway would be the most important and this was subsequently confirmed when the plant became fully operational. This pathway arises because one of the fission products, ruthenium-106, is taken up very efficiently and effectively reconcentrated by the seaweed *Porphyra umbilicalis*. At that time this was harvested on the coast close to Sellafield and was used to produce a foodstuff called *laver bread* which is consumed in South Wales. Thus the factor which controlled discharges from the plant was the dose from ^{106}Ru to the gastrointestinal tract of laver bread consumers. Later, harvesting of porphyra ceased and the emphasis shifted to the other pathways, including plutonium in winkles and caesium in fish.

The significance of deposition of activity on sand and silt is that beaches and estuaries become contaminated and this leads to external radiation exposure of people using the beaches and of fishermen, for example, when handling nets and equipment. More recently, resuspension of activity into the atmosphere when the beaches dry out has been identified as a possible exposure pathway, as has transfer from the sea via seaspray into the atmosphere and then onto land. This would be of particular importance in the case of relatively small quantities of plutonium and other actinides.

The activity taken up by fish is primarily caesium-137. Because the caesium is soluble in sea water and because of its 30 year half-life, it disperses over a very wide area giving a low but measurable concentration not only in the Irish Sea but also in the North Sea and even in the Norwegian Sea. This leads to low but measurable concentrations of the activity in fish caught over a very wide area. The dose received from this pathway by individual consumers of fish is very low, but because of the large number of consumers the total collective dose is appreciable, reaching about 200 man-Sv per year when discharges were at their peak.

Recently, the levels of discharge have been considerably reduced and the critical exposure pathway is liable to change. One important point which emerges from experience at Sellafield is that when long-lived activity such as plutonium is released into the environment it could give rise to a continuing radiological impact.

Another important example is the uptake of radioiodine, mainly ^{131}I, in the thyroid as a result of releases of fission products to the atmosphere. The uptake may be due to inhalation of airborne iodine or consumption of milk from cows which have grazed on contaminated pasture. In either case, exposed children comprise the critical group because of the relatively large intake by children in proportion to the size of the thyroid.

In all of these cases, the discharge limits are set so that the critical group receives doses well below the permissible level.

11.3   RADIOACTIVE LIQUID WASTE

Various approaches are used in the management of radioactive liquid wastes. Where the level of radioactivity is very low, liquid wastes can often be discharged either as a trade waste to sewers or directly into a river or lake, or into coastal waters. As discussed earlier, a period of storage might be worthwhile in order to allow decay of relatively short-lived radionuclides. In this case, a number of delay tanks are provided and when one is full it is isolated until the activity has decayed to an acceptable level for release. Before discharge the contents of the tank are stirred, sampled and analysed to confirm that it is within the prescribed activity limits.

Liquid wastes of higher specific activity may be treated by ion-exchange, evaporation or chemical treatment. These processes result in the concentration of the radioactivity from the liquid waste into a much smaller volume of residue or sludge which may be suitable for conversion into a solid waste for storage or disposal. The effluent from the process would normally be of sufficiently low activity to permit its discharge.

Discharge to sewers is a very convenient method where both the volume and specific activity are low but it is strictly controlled. The main considerations are that sewage, sewer walls and sewage works become contaminated, resulting in a hazard to sewage workers. In addition sewage sludge is often used as a fertilizer and, if it is contaminated, could result in contamination of crops.

Discharge of radioactivity into rivers is limited by the subsequent use of the river water. Most major European rivers are sources of drinking water and of water for irrigation of crops and watering of cattle. These considerations mean that discharges into most rivers are limited to quantities of the order of a few TBq per year, depending on local conditions. Similarly, in the case of lakes, the limited dispersion usually limits discharge rates of most nuclides to relatively low levels, again typically a few TBq per year.

Relatively large quantities of radioactivity can be discharged into the sea provided that steps are taken to ensure adequate dilution and dispersal. The limits are usually dictated by food chains involving reconcentration effects in marine organisms as described in Section 11.2. At most coastal locations, discharges of some hundreds of TBq (a few thousand curies) are feasible.

Whatever the mode of liquid discharge, authorizations are under continuous review and the safety of the procedure is confirmed by environmental monitoring programmes. In the case of marine discharges, this includes measurements on sea water, fish, seaweed and the sea bed.

11.4 RADIOACTIVE GASEOUS WASTE

Releases of gaseous or airborne particulate radioactivity to the atmosphere present more direct exposure pathways than other forms of disposal and with a few exceptions, such as the noble gases, the discharge limits are quite low. The exposure pathways include external irradiation, inhalation and ingestion by various routes (see Section 11.2). The general philosophy is to reduce the activity being released to atmosphere as far as practicable and then to release it in such a way as to obtain adequate dispersal.

The methods available in practice to achieve these objectives are:

(a) filtration to remove particulate activity,
(b) scrubbing or adsorption systems to reduce gaseous activity, and
(c) dispersal via a chimney stack.

Filtration is a very common method and is usually incorporated in extracts from any area containing significant activity. The efficiency of the filter depends on the size of particles but is usually better than 99.9 per cent. It must be remembered that the filters become radioactive and when they are changed they must be handled with care and treated as solid radioactive waste. Scrubber units and adsorption beds are bulky and expensive items and are used only in large installations such as nuclear reactor plants.

When relatively small amounts of activity are involved the releases to the atmosphere are usually from an extract discharging at, or even below, roof level. Care must be taken in the siting of such extracts since, under certain weather conditions, eddies and currents may cause the released activity to re-enter the building through air intakes or even through open windows. A chimney stack is preferable but the additional cost is only justifiable when large quantities of activity are being released. Ideally the chimney should be two to three times the height of surrounding buildings to obtain good dispersal. This is illustrated in Fig.11.2 which shows the dispersion from (a) a short stack and (b) a long stack on a large building. Even with the high stack, diffusion or spreading of the plume means that at some distance downwind the concentration of activity at ground level will show an increase. The maximum concentration usually occurs at a distance downwind of 10 to 20 times the height of the stack, depending on wind and weather.

As will liquid wastes, the adequacy of controls is confirmed by detailed environmental monitoring programmes.

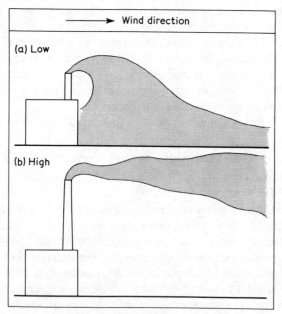

Fig. 11.2 Dispersion from chimney stacks

11.5 RADIOACTIVE SOLID WASTE

Radioactive solid wastes arise in various forms in nuclear facilities and from medical and industrial applications of radioactivity. They are usually considered to fall into three broad classes *low-*, *intermediate-* and *high-level* wastes.

Low-level waste (LLW) consists typically of general trash from contaminated areas as well as items of lightly contaminated or activated plant and equipment. Intermediate-level wastes (ILW) arise mainly in nuclear facilities and include solidified process residues and significantly activated items. The definitions of LLW and ILW, in terms of specific activity, vary from country to country but, typically, LLW would comprise waste of less than about 10^9 Bq/m^3 and ILW would be any waste, other than HLW as defined below, exceeding this value. With both LLW and ILW, disposal is often facilitated by segregation of the waste, preferably at the point of arising, for example into wastes of different specific activity, or into combustible and non-combustible material. Reduction of volume is often a useful procedure and this can be achieved either by compression of the waste or, in the case of combustible waste, by incineration. In the latter case, the flue gases from the incinerator must be filtered and the filters and ash then comprise the reduced volume of waste.

The term high-level waste (HLW) is usually reserved for the raffinate from the first extraction cycle in a fuel reprocessing plant. This stream contain 99% or more of the fission product activity and much of the alpha-emitting higher actinides. The specific activity of this waste is so high that it generates significant heat and special provision has to be made for cooling. Initially, the waste is stored as a liquid in high integrity tanks with multiple cooling systems and located inside massive concrete shielded cells. After a delay of a few years to allow some of the shorter-lived activity to decay, the waste is vitrified, which means that it is incorporated into large glass blocks in stainless steel canisters. The waste is then in a suitable form for safe storage for an extended period.

Storage can only be regarded as a temporary solution to the problem of managing radioactive waste. This is because many types of waste contain long-lived radioactivity. For example the activity in HLW includes the fission products Sr-90 and Cs-137 with half-lives of 28 and 30 years respectively. After about 500 years these will have decayed to negligible levels but significant amounts of longer-lived nuclides will remain, including the alpha-emitters americium-241 (^{241}Am, half-life 434 y), plutonium-239 (^{239}Pu, half-life 24 000 y) and neptunium-237 (^{237}Np, half-life 2·2 million years).

The need to establish safe methods of radioactive waste disposal is an important factor in the public acceptance of nuclear power. In some countries, further development of nuclear power has been made dependent on establishing the feasibility of the safe disposal of wastes.

Various methods of disposal are either in use or are under consideration, including underground disposal and sea disposal. A method practised by smaller users of radioactive materials, such as hospitals, universities and industrial users, is disposal with ordinary trade waste. The principle which is being applied here is *dilution*, the odd package containing a small quantity of radioactivity being well diluted and buried among the vast quantities of ordinary trade and domestic waste on the local refuse tip. This approach requires very strict control and the quantity per package is usually limited to about 0.1 MBq

Land burial has also been practised by the nuclear industry, but only at specially selected sites. For the future, the industry is planning two types of land disposal facilitates, or waste respositories as they are sometimes known. The first is a shallow-land facility in the form of a large engineered concrete structure 10–30 m below the surface, usually in clay beds. This type of repository is suitable mainly for LLW, but also for short-lived ILW.

The second type is a deep underground disposal facility, perhaps about 300 m or more below the surface. During the emplacement phase these would be similar to mine workings, with shafts and tunnels. The wastes would be placed in tunnels which would then be back-filled with absorbent materials and, eventually, the workings and shafts would be completely sealed. This type of repository is potentially suitable for both ILW and HLW.

The important consideration in underground disposal is that the waste containment will eventually be lost and activity will be leached into groundwater. This could result in the contamination of drinking water supplies or of crops. Disposal sites must be carefully selected to minimize these effects. Another consideration, particularly for shallow-land sites, is the possibility of human intrusion at some future time. Examples of intrusion are drilling of boreholes, tunnelling and excavation. Whilst controls over the development of the site should be able to ensure that intrusion does not occur for some hundreds of years, beyond this time it is not possible to guarantee that intrusion would not occur. For this reason, the amount of long-lived waste placed in shallow-land repositories will need to be strictly limited.

The other possible mode of disposal is to sea. This has been practised to a moderate extent by a number of countries under the auspices of an international organization, the Nuclear Energy Agency OECD. There is little doubt that, in technical terms, sea disposal is a safe and economic option for some categories of radioactive waste. Disposal of solid waste in the deep ocean results in a very long delay before the activity returns to mans environment and an enormous dilution of activity. However, in recent years some countries have opposed the practise and, at the time of writing (1986), there is a moratorium on sea disposal.

11.6 REGULATIONS

The regulations and legislation applicable in the UK are described in some detail in Chapter 14, but the important aspects relating to the disposal of radioactive waste are summarized briefly below.

The majority of small users of radioactive sources are required to register with the Department of the Environment under the provisions of the Radioactive Substances Act, 1960. Authorizations for disposal are granted by the Department after consultation with the Ministry of Agriculture, Fisheries and Food, and Local Authorities including river boards.

Operators of nuclear installations such as nuclear power stations, fuel manufacturing and reprocessing plants are exempted from the requirement to register their use and storage of radioactive materials. Instead, they are subject to the provisions of the Nuclear Installations Act 1965, 1969 and amendments. In the latter Act reference is made to the Radioactive Substances Act and the responsibility for authorizing discharges again rests with the Department of the Environment (in Wales, the Welsh Office) and the Ministry of Agriculture, Fisheries and Food. In Scotland the responsibility for authorizing discharges rests with the Scottish Development Department.

SUMMARY OF CHAPTER

Main sources of radioactive waste are reactors and the use of radioisotopes in medicine, industry and research.

Principles applied are release and dispersal, storage and disposal.

Consequences of disposal: dose to population via exposure pathways
   – direct or via food chains.

Limiting pathway is critical exposure pathway and population group
   receiving highest dose is critical group.

Liquid waste: low level liquid wastes discharged to environment. High
   specific activity wastes treated by ion-exchange, evaporation or
   chemical treatment.

Gaseous waste: release to atmosphere after filtration or scrubbing.
   Discharge from stack to achieve good dispersal.

Solid waste: LLW, ILW and HLW. Storage is only temporary solution for long-lived wastes. Segregation and volume reduction facilitate disposal. Possible routes are shallow or deep underground
   disposal and sea disposal.

Regulations: authorization required for storage or disposal of
   radioactive waste.

REVISION QUESTIONS

1. List the three general approaches for dealing with radioactive
waste. Give a practical example in each case.

2. Discuss the concept of *critical exposure pathway* and give an
example of such a pathway which involves a *food chain*.

3. What are the alternative discharge routes for low level liquid waste?
Discuss the factors which limit the quantities which may be discharged in each case.

4. What are the possible exposure pathways resulting from releases of
radioactivity to atmosphere? How would the exposure from these
pathways be limited?

5. Discuss the possible disposal routes for radioactive waste. Broadly,
what types of wastes are suitable for each route?

X-rays and radiography

12.1 INTRODUCTION

X-rays were discovered in 1895 by the German physicist Wilhelm Conrad Roentgen. During the course of experiments on electrical discharge tubes he noticed that a screen coated with barium platino-cyanide glowed when placed near the tube. After further investigation he concluded that the fluorescence was being caused by invisible rays which were capable of penetrating not only glass but also optically opaque materials. He also found that he could photograph the interior of objects. For example the bone structure of the hand could be photographed because the bone attenuates X-rays to a greater extent than the soft flesh. Photographs obtained by passing X-rays through objects (or people) in this way are known as *radiographs*.

The medical implications of this discovery were immediately appreciated and, within a few months, physicians in many parts of the world were using X-rays as an aid to diagnosis. These physicians became known as *radiologists* and, as mentioned in Chapter 5, many of them died as a result of excessive exposure to radiation. X-rays are now widely used in medicine not only for diagnosis but also for the treatment of diseases. They also have applications in industry and in research.

Like light, radio waves and γ-rays, X-rays belong to the electro-magnetic group of radiations. They have no mass or charge, but have a wavelength which depends on their energy. They differ from γ-rays in two important respects. Firstly, γ-rays originate within atomic nuclei whereas X-rays originate from changes in the electron orbits. Secondly, γ-rays from a given source have definite discrete energies but X-rays usually have a broad range or spectrum of energies up to some characteristic maximum value.

The most important method of producing X-rays depends on a process known as *bremsstrahlung* which is German for *braking radia-*

tion. Bremsstrahlung X-rays are produced when charged particles, usually electrons, moving with a very high velocity are slowed down rapidly by striking a target. For example when beta radiation from a radioactive source impinges on a shielding material, bremsstrahlung is produced. The efficiency of X-ray production by this means is very dependent on the atomic number of the target material, high-Z materials giving a much higher yield than low-Z materials. (This is the reason for using low-Z materials such as Perspex for shielding beta sources.) In any case the intensity of X-rays produced by β-particles is too low for most applications. The method used to produce X-rays for medical and industrial purposes is similar to the method used by Roentgen but modern equipment is much safer and more efficient.

12.2 X-RAY EQUIPMENT

12.2.1 General

X-rays are produced when electrons moving with high velocity are suddenly stopped by a material of high Z, and so an X-ray generator requires a source of electrons, a means of accelerating them to a high velocity and a target at which they are directed. An X-ray set consists of a tube and various electrical circuits which are usually in a separate control unit.

12.2.2 Hot cathode X-ray tubes

The modern type of X-ray tube, illustrated in Fig. 12.1, consists of a cathode and an anode inside a glass tube evacuated to an extremely

1. Bulb   4. Cathode block   6. Focus
2. Cathode   5. Anode   7. Target
3. Filament

Fig. 12.1 A typical stationary anode X-ray tube

low pressure. The cathode is the source of the electrons and consists of a tungsten filament heated to incandescence by an electric current which 'boils out' electrons. The electrons are accelerated to the target by a high voltage applied between the anode and cathode.

The target is part of the anode assembly and is constructed of a material of high Z in order to achieve the best possible efficiency of X-ray production. However, even when the efficiency is as high as practicable, less than 1% of the energy of the electrons appears as X-rays. The remainder appears as heat and so the target must have a high melting point and also be able to dissipate the heat. This is achieved by constructing the anode of copper, which has a high thermal conductivity, with a tungsten target insert facing the cathode.

The copper anode is sometimes in solid form and has a finned radiator extending outside the tube to assist cooling. In higher power sets the anode is hollow and is cooled by circulating oil or water through it. In applications such as radiography it is important, in the interests of good definition, that the source of X-rays is very small. The filament is, therefore, mounted in a concave cup which focuses the electrons on to a small area of the target. Special measures are then necessary to prevent overheating of the target and the anode may consist of a rotating disc. In that case, the effective target area is still small but the heated area is greatly increased and the tube may be heavily loaded without melting the target. This type of tube is used in medical X-ray sets, in which very high intensities and short exposure times minimize difficulties due to body movement.

12.2.3 Electrical supplies and controls

The electrical supplies required for the operation of an X-ray tube are a low voltage supply to the filament and a very high voltage to apply between anode and cathode. The supplies are usually derived from a.c. mains. In the case of the filament supply, a step down transformer is used to provide a voltage of about 12 volts a.c. at a current which can be varied up to a few amperes. The tube voltage is provided by a high voltage transformer which steps up the mains voltage (250 V a.c. in UK) to the level required for operation of the tube. This is normally in the range of 5000 V up to some millions of volts depending on the application. Since the voltage is derived from a.c. mains, the voltage between the anode and cathode is also alternating. A typical power supply is illustrated in Fig.12.2.

1. Tube voltage control   5. Filament current control
2. Voltmeter   6. Low voltage transformer
3. High voltage transformer 7. Filament current ammeter
4. Tube current ammeter   8. X-ray tube

Fig. 12.2   Simplified power supply for X-ray set

In a hot cathode X-ray tube electrons can flow only from the cathode to the anode since there is no source of electrons at the anode. However, the electrons will flow to the anode only when attracted by a positive voltage (remember that electrons are negatively charged) and so the tube will draw current and emit X-rays only when there is a positive voltage on the anode. The variation of the anode voltage with time is shown in Fig. 12.3. The shaded areas represent the periods when electrons are flowing to the anode (i.e. current is flowing) and producing X-rays at the target. The X-rays are not, therefore, produced continuously but as a series of pulses. If, for example, the a.c. supply is of 50 cycles per second, then 50 pulses of X-rays are emitted per second. This pulsing is of some importance is designing X-ray monitoring equipment since some types of detector may give a reading which depends on the frequency of X-ray pulses rather than on the average intensity. In some sets, special circuits are provided

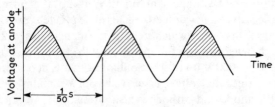

Fig. 12.3   X-ray production in a self-rectifying tube

which smooth out the X-ray production to give a more uniform output with time.

12.3 QUALITY AND INTENSITY OF X-RAYS

The quality or energy of X-rays depends on the peak voltage applied to the anode of the tube. If the peak voltage in Fig.12.3 is 200 000 V, this is expressed as 200 kV peak or 200 kVp. The maximum energy of X-rays produced is 200 kV but only a very small fraction will have this value and most of the X-rays will be of lower energy. The quality of the X-rays is, however, defined in terms of this peak energy and they are said to be 200 kVp X-rays. The penetrating power of X-rays is highly dependent on their energy. For example, the quality of X-rays used to radiograph a person's hand would be much too low to radiograph a 10 mm steel plate. The voltage on the tube is therefore set to give the appropriate quality of X-rays for each application. A few examples of suitable operating voltages and exposures for medical and industrial radiographic applications are given in Table 12.1

Whilst the voltage on the tube controls the quality of X-rays, the dose rate is governed also by the current flowing in the tube, that is between the anode and cathode. This current, expressed in milliamperes (mA), is limited by the number of electrons ejected from the cathode. This is controlled by the temperature of the filament which in turn depends on the current flowing in the low voltage filament circuit. Since the dose rate is controlled by the tube current, the total dose in a particular case depends on the tube current multiplied by the time of exposure. The same dose would be received from an exposure of 10 mA for 1 s as for 1 mA for 10 s. In both cases the exposure is 10 milliampere-seconds or 10 mA s.

Table 12.1 Typical operating voltages for radiography

		kVp	Dist (m)	mA s
Medical diagnosis	Finger bone	40	1	10
	Skull	80	1	100
	Chest	80	2	80
	Pelvis	120	1	150
Industrial	6 mm steel	120	0·5	10
	25 mm steel	160	0.5	200

The dose rate from an X-ray set is very high compared to dose rates from typical sealed gamma sources. The output is usually expressed in terms of the absorbed dose rate in m Gy per minute at 1 m from the set for a tube current of 1 mA. Some typical outputs are shown in Table 12.2.

Table 12.2   Typical output of X-ray sets

Equipment and filtration	m Gy/min/mA at 1 m
50 kVp beryllium window tube	100
100 kVp 3 mm Al (external to tube)	30
200 kVp 2 mm Cu + 1 mm Al (external to tube)	20
300 kVp 3 mm Cu + 1 mm Al (external to tube)	10
500 kVp 3 mm Cu + 1 mm Al (external to tube)	25

The significance of the beryllium window mentioned in Table 12.2 is that at low voltages the penetrating power of X-rays is so low that a high proportion would be absorbed by a glass bulb. The use of a thin beryllium window minimizes this loss of output.

At higher voltages additional absorbers, or filters as they are known, are provided in the X-ray beam. It has been mentioned that X-rays of all energies up to the peak voltage are produced. Only the small fraction with the higher energies are useful, the remainder are undesirable in many cases. For example, in medical radiography, the low energy fraction would not contribute to the radiograph, but would result in unnecessary skin dose to the patient. The use of filters, usually an appropriate thickness of aluminium, selectively absorbs the low energy or soft radiation without significantly affecting the useful beam.

12.4   PROTECTION AGAINST X-RAYS

12.4.1   General Principles

Unlike radioisotopes, which emit radiation continuously, X-ray sets can be switched on or off at will. During operation the dose rate from the set may be very much higher than from small sealed sources. The equipment must be run in such a way that the operator does not expose any part of his body to the direct beam and no other person is accidentally exposed. The general principles which are applied to the protection of personnel are as follows:

(a) Adequate training of all personnel who operate or use X-ray equipment in the correct operating procedures and in the hazards involved.

(b) Limitation of the beam size to the minimum necessary by provision of shielding and collimators built into the set.

(c) The use of suitable filtration to remove unwanted soft radiation.

(d) Operation of X-ray equipment in a shielded room whenever possible. The controls are situated outside the room and an interlock is provided which prevents operation of the set whilst the door is open.

(e) Provision of automatic visible and audible warnings that the set is operating or about to operate.

(f) Confirmation of the effectiveness of control by means of a system of personal and area monitoring.

The measures applied in any given case depend very much on the type of work and the local conditions. In medical applications, for example, the reassurance and well-being of the patient is of great importance and has a bearing on the protective measures taken by the operator. Protection against radiation in medical applications is described in greater detail in Chapter 13. The following sections are concerned with protection against X-rays in industrial applications and in research.

12.4.2 Protection in industrial radiography

Radiography is a very important method of non-destructive testing. It is widely used for the examination of castings, forgings and welds to detect internal defects or weaknesses. Sometimes radiography is part of a production process and is properly set up in a shielded facility with the appropriate safeguards. In other cases, it is carried out in conditions which are far from ideal, such as on a construction site.

In the UK, factories and sites in which radiography is used are usually subject to the provisions of the Ionizing Radiations Regulations, 1985, under the Health and Safety at Work Act and observance of the regulations is a statutory obligation.

The general principles which apply to the control of hazards from radiography are as follows:

(a) Whenever practicable, radiography or any other process utilizing ionizing radiation from machines or sealed sources should be carried out within a shielded enclosure.

(b) The control panel for the apparatus is to be located outside the enclosure and devices should be provided to ensure that if any door of the enclosure is opened while the apparatus is energized, the apparatus is automatically de-energized.

(c) For the protection of persons accidentally shut inside the enclosure, a means of communication is required to enable them to summon help. In addition, one or more of the following facilities should be provided for such persons:
    means of exit,
    means of de-energizing the apparatus, or
    a shielded area.

(d) Audible or visible signals (or both) are required to be given when the apparatus is about to be energized, and a different signal whilst the apparatus is energized.

Where it is impracticable to carry out the work in a special enclosure, effective measures are required to exclude unauthorized personnel from a suitably marked area around the work.

The UK regulatory position is discussed in greater detail in Chapter 14.

12.4.3 Protection in research applications

The two main applications of X-rays in scientific research are X-ray crystallography and X-ray spectrometry. Crystals are regular arrangements of atoms and it has been found that definite patterns of scattered radiation, known as diffraction patterns, result from irradiating crystals with X-rays. The precise nature of the pattern gives important information regarding the structure of the crystals.

In crystallography, very high intensity X-ray beams of small cross section are used, so posing special health physics problems. In X-ray spectrometry substances are irradiated by X-rays and, as a result of absorbing energy, are excited and emit secondary X-rays. The energy of this secondary radiation is characteristic of the element and so measurement of the secondary X-ray energies enables analysis of the substance.

The precautions to be observed in this type of work include:

(*a*) The apparatus should be adequately shielded and where access to the inside of the apparatus is necessary, either the machine must be automatically de-energized or effective steps must be taken to prevent the insertion of any part of the body into the beam.

(*b*) Where a camera or slit collimating system is in use the useful beam should be fully enclosed so as to provide adequate shielding.

(*c*) Automatic visible or audible warning devices should operate whenever the apparatus is energized.

12.5   MONITORING OF X-RAY INSTALLATIONS

An important part of the commissioning procedure of any X-ray installation, or other facility which produces ionizing radiation, is a thorough radiation survey. Particular attention is paid to possible weaknesses in shielding, such as joints in the shielding material, viewing windows, doors and holes or ducts for services. The survey, which is usually at maximum tube voltage and current, is made under normal operating modes and then under other possible operating modes. Consider, for example, the case of a facility in which the X-ray beam is intended to operate in the horizontal plane and adjacent areas are shielded by thick walls. If the orientation of the set is changed and the beam operated in a vertical direction, would unacceptable levels of radiation occur in the areas above or below the facility? It should be borne in mind that if such a change is possible it is quite likely that, one day, it will be made. If it is found that excessive radiation levels could possibly occur in adjacent areas, measures must be taken to prevent, or at least to give warning of, the situation. This can be done by mechanically preventing the beam being operated outside set limits, by provision of additional shielding or by the installation of area radiation monitoring equipment with warning signals. In general the two former methods are preferable.

Clearly, questions of this sort should be considered at the design stage but it is essential to confirm the safety of the facility by direct measurement. Surveys are repeated periodically, particularly when any changes in operating procedure are introduced.

Care is necessary in the selection of instruments for monitoring X-rays. The major problem is that of energy response. Many instruments which are satisfactory for γ-ray and higher energy X-ray work seriously underestimate the dose rate from X-rays of below about 100 kVp. For low energy work, instruments incorporating thin window ionization chambers are probably the most suitable although they are sometimes lacking in sensitivity. Another possible problem when pulse-type instruments (e.g G-M tubes) are used is that the instrument may saturate in high X-ray dose rates and yet still appear to be working satisfactorily. This is due to the pulsed nature of the X-rays which allows the instrument to recover between pulses. The instrument might then record the X-ray pulse rate rather than the average dose rate. This problem is fortunately rare in equipment of modern design but may arise as a result of a fault.

The safety of a facility is ultimately judged by the radiation doses received by operators and other persons working in the vicinity. These are normally measured by film badges though some establishments now use thermoluminescent dosemeters. It is often worth while using a few film badges or other dosemeters to monitor fixed locations around the area on a routine basis. It should be borne in mind that personal monitors are small in area. X-ray beams, particularly in crystallography, may also be small in cross section. It is quite possible for a beam to miss a personal monitor but nevertheless to irradiate the worker.

SUMMARY OF CHAPTER

X-rays: electromagnetic radiation – originate from changes in electron energy levels or as bremsstrahlung – electrons striking high-Z target material.

Equipment: tube and control unit.

Tube: cathode and anode in evacuated glass tube.

Controls: high voltage controls quality of X-rays, low voltage filament supply affects intensity.

Protection: set can be switched off – but hazardous when operating. Safety depends on a combination of staff training, shielded rooms, filtration, warning systems, personal and area monitoring.

REVISION QUESTIONS

1. Describe, with the aid of a sketch, the operation of a hot cathode X-ray tube.

2. Compare the effect of varying the tube voltage and the tube current on the radiation output.

3. Compare the radiological hazards posed by large γ-sources and X-ray sets.

4. Design an enclosure for the routine X-radiographic examination of large metal castings. Indicate the safety features which are included.

5. What special problems arise in the monitoring of X-rays?

13

Radiation protection in medicine

13.1 APPLICATIONS

Ionizing radiation is a powerful tool in many branches of medicine, both as an aid to diagnosis and a means of therapy (treatment). The most familiar application is the diagnostic X-ray, which is used in a wide variety of examinations. Radiation can induce cancer and yet, curiously, it can also cure the disease in some cases. This is because cells which are dividing rapidly are particularly sensitive to radiation and since cancers are groups of cells dividing in an uncontrolled manner, it follows that they are often more sensitive to radiation than normal cells. There are various other conditions which can be treated by radiation, notably skin diseases, but because of the hazard involved, other methods are usually tried first. The most common method of treatment is by X-rays but γ-rays from sealed radium, cobalt-60 or caesium-137 sources are often used. Comparatively small sealed sources may be surgically implanted in the region requiring treatment. Sealed β-ray sources such as strontium/yttrium-90 are sometimes used to irradiate the skin or eyes. Another method in the large specialist hospitals is the use of cyclotrons, Van de Graaff machines and linear accelarators which produce beams of electrons or neutrons.

Unsealed sources of radioactivity also have applications in diagnosis and therapy. The administration of a radioactive solution either intravenously or orally results in the uptake of the activity into various parts of the body. Depending upon the particular radionuclide used, and the form in which it is administered, the radioactivity may be distributed fairly uniformly over the whole of the body, or it may concentrate in specific organs. In therapeutic applications the object is to deliver a pre-determined dose to some particular organ whilst minimizing the dose to the rest of the body. Diagnostic tests consist of observing the behaviour of the administered radionuclide either by measurement of excreted radioactivity or by means of radiation detectors outside the body. For example, the functioning of the thyroid gland can be tested by oral administration of iodine-131 solution and monitoring the rate at which the iodine is taken up by the thyroid.

A technique which has been developed in recent years and which is of great importance in medical diagnosis is *computed tomography*, popularly known as *scanning*. A tomograph is an image of a section, i.e. a slice, through an object, in this case the human body. There are two different approaches, known as transmission tomography and emission tomography and the application of both systems is briefly described within this chapter.

13.2 GENERAL PRINCIPLES AND ORGANIZATION

Special problems arise in radiation protection in medicine because the well-being and reassurance of the patient is of prime importance. A patient who has been given a large intake of radioactivity can represent a significant radiation hazard to staff, other patients and visitors. The normal methods of protection described in earlier chapters such as shielding, distance and containment cannot be applied in the usual way. However, with common sense, the patient can be adequately cared for without excessive hazard to others.

The basic principles of radiation protection in medical applications are as follows:

(a) Radiological examination or treatment should be undertaken only when it offers advantages over other methods.

(b) Where possible any such examination or treatment should take place in a special radiological department or in special wards.

(c) The dose to the patient should be minimized by using the best available techniques and measures should be taken to reduce as far as possible the dose to other parts of the body.

(d) Routine measures should be taken to minimize gonad dose, for example in the case of X-rays, by limitation of the field size and by provision of shielding.

(e) Special consideration should always be given before exposing pregnant women or young children to radiation.

(f) All radiological procedures should be performed in such a way as to minimize the dose to other persons.

The organization and responsibility for radiation protection in medical establishments varies from country to country. In the UK, medical uses of radiation are subject to the requirements of the Ionizing Radiation Regulations 1985, under the Health and Safety at Work Act, see Chapter 14. The responsibility for radiation protection in hospitals lies with the Controlling Authority, usually the Hospital Management Committee. Their authority is exercised via a Radiological Safety Committee and a Radiation Protection Adviser (RPA). Since the RPA may cover many departments, or even several hospitals, it is usually necessary to appoint a Radiological Protection Supervisor (RPS) on a part-time basis in each department.

It is essential that all staff who are liable to be exposed to radiation in the course of their work are properly instructed about the nature of the hazard and the precautions to be taken. The hazards and methods of protection depend on the type of source and can be conveniently divided into two types.

(*a*) protection against sealed sources, and
(*b*) protection against unsealed radioactive substances.

Sealed sources in this context refers not only to encapsulated radionuclides emitting β- or γ-rays but also to machines which generate radiation, including X-ray sets, electron accelerators and neutron generators. Essentially the distinction between sealed and unsealed sources is that in the former case, barring accidents, there is no radioactive contamination problem.

13.3   PROTECTION AGAINST SEALED SOURCES

13.3.1   Diagnostic radiography

X-ray sets are the main type of sealed source used for diagnostic purposes and a number of techniques are available to suit different applications. The most familiar technique is exactly as used in industrial radiography, that is by placing the part of the body to be examined between an X-ray set and a photographic plate. Careful selection of the X-ray quality (voltage) and the type of photographic emulsion enables good quality radiographs to be obtained with quite low doses to the patient. For example, using the best techniques available a chest X-ray can deliver as little as 100 μSv (10 mrem) to the chest of the patient, although typically the dose is about 1000 μSv

(100 mrem). The dose to the gonads is much lower than these values provided that the beam size has been properly adjusted so that the gonads are outside the primary beam. In many cases it is necessary to take more than one radiograph, but clearly the number of 'shots' should be minimized.

The dose to the radiographer is minimized by good design of the facility, for example by the provision of a shielded cubicle in which the radiographer must stand in order to operate the set. Occasionally difficulties arise; for example, young children may need to be held in the correct position. If a harnesss cannot be used it is better for the parent to hold the child rather than the radiographer, since the parent is unlikely to be exposed frequently in this way. A similar problem sometimes occurs in dental radiography when it is not possible to clamp the film in position in the mouth. In this case the patient should hold the film rather than the dentist or his staff.

An important point to bear in mind in medical X-ray work is that a significant reduction in dose can be obtained with quite thin shielding because of the relatively low X-ray energy used (often less than 100 kVp). For example, lead impregnated materials are available which can be made into aprons and gloves and are equivalent in shielding ability to a millimetre or so of lead.

In addition to the hazard from the primary beam, X-rays may be scattered from the patient or near-by materials, so constituting a further hazard.

13.3.2 Diagnostic fluoroscopy

In fluoroscopy, the photographic plate used in radiography is replaced by a fluorescent screen. The screen fluoresces under irradiation and therefore gives a live picture. The method is used in two ways, namely mass miniature radiography and fluoroscopic examination. In mass miniature radiography the patient stands in front of the screen in the usual way but the fluorescent screen is photographed using a small film. When a complete roll of film has been exposed the film is processed and the radiographs are viewed on a projection screen. The main advantages are economic since the costs of film, processing and storage are greatly reduced. The disadvantage is that the dose received by the patient is about 5 mSv compared to less than 1 mSv for a conventional radiographic chest examination. (These are the doses at the surface of the body.)

In fluoroscopic examinations, the radiologist stands behind the screen and views the picture presented by the fluorescence of the screen (see Fig. 13.1). The essential feature of this technique is that the beam can remain on for 30 seconds or more so that a moving picture is presented to the radiologist (e.g. as the patient breathes). The dose received by the patient is rather high, typical dose rates are 10 to 20 mSv per minute. The method has sometimes been used in order to avoid film processing. In view of the higher dose involved this is no longer regarded as an adequate reason for fluoroscopy and it has been largely supplanted by other techniques.

Great care must be exercised in the design and operation of a fluoroscopic facility if the radiologist is to avoid excessive exposure. He relies on the thickness of the screen, a lead apron and a lead sheet hanging below the screen. It should be possible to switch the beam on and off from behind the screen, a foot-switch is usually used. A timer which automatically switches off the beam after, say 20 seconds helps to avoid excessive exposure.

In hospitals, fluoroscopy is generally used in conjunction with a technique known as image itensification, in which the image on the fluorescent screen is intensified by electronic means. This permits a reduction in dose to both the patient and the radiologist. The method may be further refined by using a closed circuit television system which enables the radiologist to be in a separate room or in a shielded cubicle. Obviously the cost of the equipment is greatly increased.

Fig. 13.1 Fluoroscopic examination

13.3.3 Transmission computed tomography

As was noted earlier, tomography, or scanning, is an important tool in medical diagnosis. One technique is known as *transmission computed tomography* and involves passing a beam of X-rays through the body and measuring the amount transmitted using an array of collimated (i.e. directional) detectors. The principle of operation is illustrated in Fig. 13.2 which shows a system in which the source and detector system are rotated around the body. This gives a very complex pattern of detector response which can be analysed by computer to give an image of the section which has been scanned.

The system reveals physical or anatomical features within the body and so can detect the presence of an abnormality such as a tumour.

The dose received by the patient depends on the type and extent of the examination and would, typically be about 10 mSv (1 rem), but only locally. The equipment is normally located in a shielded room and only under exceptional circumstances should the radiographer need to enter the room during the scan.

13.3.4 Radiotherapy

The main application of radiotherapy is in the treatment of cancer. The aim is to deliver as high a dose as possible to the malignant tissue

Fig. 13.2 Schematic illustration of transmission computed tomography

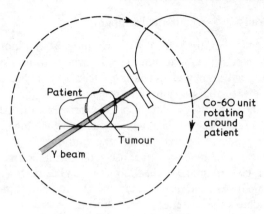

Fig. 13.3   Treatment of brain tumour using ^{60}Co teletherapy unit

without causing excessive injury to surrounding healthy tissue. Absorbed doses of a few tens of Gy are required and they are usually delivered as a series of smaller doses, for example 20 doses of 2 Gy at intervals of 2 or 3 days. This fractionation is necessary to reduce side effects.

A common method of treatment is by X-rays of about 200 kVp but when the tumour is some distance below the surface this results in a skin dose higher than that to the tumour. In such situations it is preferable to use more penetrating radiation such as high energy X-rays of about 6 MVp or γ-rays from a large cobalt-60 source. This also serves to reduce the dose to bone.

In addition to selection of the appropriate energy, the dose to healthy tissue is minimized by varying the direction of the beam through the body. This is done either by using a different orientation for each treatment, or, with more sophisticated equipment, by continuously rotating the source around the tumour during the treatment. This is illustrated in Fig. 13.3 which shows treatment of a brain tumour using a rotating teletherapy unit containing a large cobalt-60 source. Although the tumour is being irradiated continuously, the surrounding regions are exposed for only a small fraction of the time. It is essential to use a well defined beam and this is achieved by means of cones and diaphragms.

Except in some low voltage (< 100 kVp) X-ray therapy, the problem of providing local shielding is such that the treatment must be performed in a shielded room with interlocks arranged to shut down the equipment should the door be opened. A shielded viewing window and a means of communication with the patient are required. Special procedures are required to avoid accidental over-exposure of the patient. These include rigid observance of an approved operating drill, automatic timing of exposures and regular calibration of equipment.

Radiotherapy can also be effected by means of small sealed sources applied directly to the body surface, into orifices or surgically implanted in the body. The most common types of source are Ir-192 and Cs-137. Staff using the sources are protected by the use of suitable handling jigs and tools, and by sensible application of the principles of time, distance and shielding. The sources must be tested frequently for leakage and written emergency procedures should describe the actions to be taken in the event of damage to or loss of a source.

A patient containing a source may be a radiation hazard to others and this should be indicated by attaching a notice to his bed and displaying the appropriate symbol. It is preferable to use small wards and to keep the beds well spaced. The Radiation Protection Supervisor, after monitoring the patient, advises nursing staff of any time limitations on nursing procedures or visiting periods. It is possible in some cases for patients containing sources to be discharged. This is decided in individual cases after consideration of the nuclide, the half-life, and the dose rate, that is to say, the potential hazard to other persons.

13.4 PROTECTION AGAINST UNSEALED SOURCES

13.4.1 General

The laboratories in which unsealed sources are prepared and dispensed for medical use are conventional radiochemical laboratories and protection is achieved by the methods described in Chapter 8. That is by minimizing the quantities of radioactivity handled, by containing it whenever possible and by the use of good procedures and facilities. The grade of the laboratory should be appropriate to the radiotoxicity and the quantity of the nuclides in use. It will be recalled that in such laboratories special attention is paid to surface finishes and to ventilation. Fumehoods are an essential feature even if quite low levels are being used. For high level work glove boxes may be necessary. Benches, fumehoods and glove boxes should be of adequate construction to support temporary or permanent shielding which may be required if large quantities of γ-emitters are being handled.

The development of radionuclide generators, particularly Tc-99m generators, is an important step forward in minimizing contamination. The generator typically contains 0.04 TBq (1 Ci) of Mo-99 which has a half-life of 66 h and which decays to the pure γ-emitter Tc-99m which has a half-life of 6 h. The Mo-99 is absorbed onto tin dioxide and as the Tc-99m daughter is produced it is released into saline solution in the generator. The saline solution containing the Tc-99m is eluted into phials in preparation for administration.

As in all active areas, a high standard of housekeeping is essential and the usual ban on eating, smoking, drinking and mouth operations such as pipetting is applied. Protective clothing and washing and monitoring facilities must be provided. It is usually necessary to segregate high level and low level work into different laboratories since low level diagnostic tests can be ruined by cross-contamination from highly active equipment used in therapeutic work.

13.4.2 Diagnostic radioisotope tests

Radioactive tracer methods are useful in many medical examinations. The organs which can be studied by this technique include the lungs, brain, liver, spleen, kidneys, thyroid, bone and blood.

The main radionuclide used for tracer studies is Tc-99m which can be used to label drugs or compounds which behave in characteristic ways in the organ being studied. The radioiodines I-131 and I-123m are also used, particularly for tests of thyroid function. The behaviour of the tracer in the body is studied by external detectors. One example of this is the use of a simple detector held against the body, as shown in Fig. 13.4 for a test of radioiodine uptake in the thyroid. The other approach to studying the behaviour of the administered radioactivity is to use *emission computed tomography*, see Fig. 13.5. In this example of tomography, the collimated detectors measure the radiation being emitted from the radioactivity in the body and a picture can be built up of its distribution and of how the distribution changes with time. This gives information on the functioning of the organ being studied.

The dose received by the patient is usually about 1 mSv (100 mrem) for tests using Tc-99m or I-131 but can be up to about 50 mSv (5 rem) to particular organs. The dose to medical staff is mainly finger tip dose due to handling and administration of the radioactive tracer.

Fig. 13.4 Thyroid radioiodine uptake test

Under most circumstances the patient can be discharged immediately after the examination has been completed since the low activities involved do not represent a significant hazard to other people. The radioactivity is normally reduced to a very low level within a few weeks by radioactive decay and excretion.

13.4.3 Radioisotope therapy

In some circumstances radiation therapy is best performed by the ingestion or injection of radionuclide solutions into the body. The nuclide chosen is one which concentrates in the organ requiring

Fig. 13.5 Emission computed tomography system (courtesy of IGE Medical Systems Ltd)

treatment thus minimizing the dose to the rest of the body. Nuclides of fairly short half-life (a few days) are used and the quantity is selected so that the required dose is delivered from the time of administration until the nuclide decays or is excreted. The main application for radioisotope therapy is the treatment of thyroid tumours, using I-131. Quantities of up to about 5000 MBq (~ 0.1 Ci) are applied giving a thyroid dose of up to 100 Gy (10 000 rads), and a dose to the whole body of up to 1 Gy (100 rads).

Patients containing therapeutic quantities of radioactivity should be nursed under conditions which permit easy containment of radioactivity in case of contamination. Ideally special wards should be available in which all surfaces are designed to permit easy cleaning and filtered extract systems are provided. Protective gowns and gloves are worn when handling the patient, contaminated linen or excreta and a special storage area is provided for contaminated linen waste and samples of excreta. The Radiation Protection Supervisor specifies any limitations on time allowable for nursing procedures or visiting periods. Washing and monitoring facilities are provided for use when leaving the area and regular radiation and contamination surveys are made of the ward.

13.5   CONTROL AND DISPOSAL OF RADIOACTIVE MATERIALS

A large hospital may hold a very large number of radioactive sources, both sealed and unsealed, and so requires a special storage area located in a position which minimizes the risk of fire or flood damage. The importance of keeping good records of the location of each source and the regular mustering of all sources cannot be over-emphasized. Leakage tests are required to be performed on all sealed sources at least annually and any source showing significant leakage must be withdrawn from use immediately. Sources not in the main storage area should be kept and transported only in approved containers. These containers should be constructed so as to provide adequate shielding and adequate containment to prevent dispersal in the event of damage to the source. They should be clearly marked *radioactive* and carry the standard trefoil symbol. Written procedures should be available detailing the actions to be taken in the event of loss or breakage of a source.

In the UK, hospitals belonging to the National Health Service are exempted from the requirement in the Radioactive Substances Act, 1960, to register their use of radioactive materials. However, unless the quantities involved are very small, they are required to obtain an authorization to accumulate and dispose of radioactive waste. The general policy on radioactive waste disposal from hospitals is to use conventional local methods whenever possible. Low level solid waste is removed with ordinary refuse and usually finds its way to the local refuse tip. Low level liquid wastes such as laboratory washings or excreta from patients are discharged to the normal sewage system. Since most of the unsealed radioactivity used in medical work is short lived, it is usually practicable to store high level waste until it decays to an acceptable level for discharge to the sewage system. High level waste of long half-life, whether solid or liquid, can, by prior arrangement, be collected by or sent to the National Disposal Service.

Records must be kept of all waste discharges, both solid and liquid. It is not usually practicable to measure the activity and composition of the waste but sensible estimates can be made from a knowledge of the quantities of the various nuclides in use.

The disposal of radioactive waste is described in greater detail in Chapter 11.

SUMMARY OF CHAPTER

Radiation used in diagnosis and treatment: used only when it offers clear advantage over other methods.

Computed tomography: transmission or emission.

Diagnostic radiology: usually X-ray machine.

Therapeutic radiology: X-rays, γ-rays, electrons or neutrons.

Diagnostic radioisotope tests: nuclide depends on organ being studied – usually has a short half-life – low activity.

Radioisotope therapy: relatively high activity – patient may be hazard to staff and visitors.

Principles of protection: minimize dose to patient, avoid excessive dose to organs other than that being studied or treated. Minimize dose to staff and visitors.

Disposal of waste: conventional methods, i.e. sewers and local refuse collection – subject to authorization.

REVISION QUESTIONS

1. What are the main applications of radioactive substances and other sources of ionizing radiation in medicine?

2. List some basic principles which are applied in the protection of patients and staff.

3. Discuss the various methods of diagnostic radiology which are available using sealed sources and compare their merits.

4. Describe the special radiological problems which may arise in the nursing of patients containing therapeutic quantities of radioactivity.

14

Legislation and other regulations related to radiological protection

14.1 INTRODUCTION

The legislation and regulations dealing with radiological protection vary quite considerably from one country to another and are often intimately linked with other legislation, dealing, for instance, with the control of nuclear energy etc. It is not possible, therefore, to give an account of all the relevant national regulations in this book and readers who require further information on the regulatory position in a particular country should consult the bibliography. There are, however, a number of similarities in the radiological protection legislation of many countries which may best be illustrated by considering a specific example. We have chosen to discuss UK legislation in detail to illustrate how the legislation has developed to cope with different aspects of the control of hazards to workers and the public from ionizing radiations. Later sections of the chapter outline the role of international organizations in setting rules and guidelines concerning radiological protection. Finally, a very brief account is given of the more important aspects of the regulatory position in France, Germany, Japan and the USA.

14.2 THE COMMON LAW

Prior to 1942 there was no legislation in the United Kingdom directly related to the hazards from radioactive substances. The public was protected to a certain extent by Common Law, that is, the law of precedent. Some guidelines could be deduced from previous cases which did not involve radioactive substances, for example:

(a) Manufacturers, and other persons in charge of goods which they know to be dangerous are strictly liable for these goods. They must make the goods as safe as possible and explain their dangers before transferring them to another person.

(b) A person who keeps anything on his or her property which is likely to do mischief if it escapes is answerable for any damage which results from its escape.

(c) An employer is liable to pay damages to an employee if the employee is injured as a result of the employer's negligence. In general, an employer will be held negligent if he or she fails to provide safe means of working or adequate supervision and an employee is injured as a result.

(d) An employer is generally liable for damage caused by an employee. The employee may also be liable.

14.3 HISTORY OF LEGISLATION

While common law provided some safeguards, it did not lay down any positive controls over the manner in which operations involving radioactive substances should be carried out. The need for such regulations was first encountered in the luminizing industry where radium-bearing paints were used to luminize such items as watch and clock faces, etc. The controls in this industry were not very good and so the *Luminizing Regulations* were introduced in 1942. These regulations laid down certain basic rules for the luminizing industry, such as the provision of washing-up facilities, house rules for active areas and so on.

In 1946 the *Atomic Energy Act* was enacted which empowered the appropriate Minister (now the Secretary of State for Energy) to promote and control the development of atomic energy. This was followed in 1948 by the *Radioactive Substances Act* which provided for the control of radioactive substances and certain types of apparatus producing radiation, from the health and safety point of view. Under the terms of the Act certain 'Appropriate Ministers' were designated who had responsibilities in various fields. For example, the Secretary of State for Employment is designated the appropriate Minister for making safety regulations in laboratories, etc., and the Secretary of State for Transport the appropriate Minister for making transport regulations.

The Radioactive Substances Act (1960) is administered by the Radiochemical Inspectorate of the Department of the Environment in England, HM Pollution Inspectorate in Scotland and the Welsh Office in Wales. It was introduced to control the disposal and accumulation of radioactive waste. The Act provides for the registration of users of radioactive material and mobile radioactive apparatus and prohibits the accumulation or disposal of radioactive waste except in accordance with an authorization granted by the Secretary of State. Other parts of the Act deal with disposal facilities and the inspection of premises. The United Kingdom Atomic Energy Authority is exempted from the provisions of the Act, as are establishments which are subject to the Nuclear Installations Act.

14.3.1 The Nuclear Installations Acts, 1965 and 1969

The Nuclear Installations Act, originally introduced in 1959, was re-enacted in 1965 and further amended in 1969. These Acts are administered by the Health and Safety Executive (see Section 14.3.3) through the Nuclear Installations Inspectorate. They apply to nuclear reactors, fuel manufacture and processing plants, certain types of radioactive waste storage facilities and certain subcritical assemblies and radioisotope manufacturing facilities, other than facilities operated by the Crown and the UKAEA. The provisions of these Acts cover the licensing and insurance of radioactive materials and are in line with various International Conventions on Third Party Liability in the Field of Nuclear Energy (Paris 1960, Brussels 1963 and Vienna 1963). Some of the main provisions are:

(a) No person, other than the UKAEA, may operate a nuclear reactor or other installation covered by the Act except under a licence granted by the Health and Safety Executive.

(b) No person, other than the UKAEA, may operate an extraction or fuel enrichment plant except in accordance with a permit.

(c) Licensees and the UKAEA have an absolute duty to ensure that no occurrences involving nuclear matter on the site or in transport cause injury or damage to property. The liability for radioactive material arising on a licensed site remains with the licensed site until it is accepted by another licensed site. This means that extreme care must be taken when dispatching active or contaminated components off-site.

(d) Licensees are required to ensure that funds are available to meet claims up to £20 million in respect of each site, apart from research reactors for which the limit of liability is £5 million. Parliament is responsible for providing money to meet claims which exceed licensee's available funds in accordance with the Paris and Brussels conventions.

14.3.2 Radiological Protection Act (1970)

The main purpose of this Act was to establish a single national point of authoritative reference in the field of radiological protection. Its main provisions are:

(a) The setting up of a National Radiological Protection Board (NRPB) and an Advisory Committee.

(b) The functions of the NRPB are to advance the acquisition of knowledge about the protection of mankind against radiation hazards and to provide advice and information on that subject. It is empowered to provide technical services and to charge for any of its services. The services provided by the NRPB include both external and internal dose assessments, recording and analysis of doses to workers and assessment of the sources of exposure of the UK population as a whole. It also carries out research in various fields related to radiation protection and accidental releases from nuclear facilities.

(c) The duties of the Advisory Committee are to advise the Board on the practical aspects of applying current knowledge to the control of radiation hazards and also to bring to the Board's notice any international agreement or internationally recommended standards relevant to radiation protection. The Committee is required to draw the attention of the Board to any other matter requiring their advice or services, or which it considers should be the subject of further research.

14.3.3 Health and Safety at Work etc. Act (1974)

This major piece of legislation was enacted with the aim of rationalizing the existing, rather fragmentary legislation concerning the health and safety of persons at work, and of other persons who may be put at some risk by the activities of persons at work. The Act provided for the setting up of a Health and Safety Commission (HSC) which consists of representatives of employers' organizations, trades' unions and other organizations, such as local authorities. The HSC is responsible under the Act for formulating regulations, disseminating information to employers, employees and others, making arrangements for carrying out any necessary research and training, and generally providing advice on the interpretation of the Act. The Commission reports to the particular Secretary of State who is responsible for the activity or work area in question. The actual administration of the statutory requirements of the Act is the responsibility of the Health and Safety Executive (HSE) which comprises three persons appointed by the Commission. It is the duty of the Executive to make adequate arrangements for the enforcement of the relevant statutory provisions, except where such enforcement is transferred to some other authority (e.g. local authorities). The HSE enforces the various regulations through a number of Inspectorates which have been incorporated into the overall ogranization, e.g. the Factories Inspectorate, the Nuclear Installations Inspectorate etc. The Health and Safety at Work Act is a very extensive and complicated piece of legislation and persons with responsibilities subject to the Act are advised to read the detailed regulations. In Section 14.4 the relevant provisions concerning radiological protection are discussed.

14.3.4 The Euratom Directive

When the UK joined the European Communities it became subject to the provisions of the Treaty establishing the European Atomic Energy Community (Euratom). A requirement of this Treaty is that basic standards should be laid down in the Community for the protection of workers and the public against the dangers arising from ionizing radiations. These standards are promulgated via a 'Directive', the contents of which must be given effect in the member states.

The revised Directive, adopted on 1 June 1976, reflected the recommendations of ICRP as given in their Publication 9. The date for compliance with the terms of this Directive was June 1978 but in the intervening period ICRP Publication 26 was issued which contained a number of revised recommendations based on a somewhat altered philosophy of radiation protection. To take account of these new recommendations a revised Directive was adopted by the EEC Council of Ministers on 15 July 1980.

Although compliance with the revised Directive was initially set for December 1982 it was later changed to 1 April 1986 to give Member States enough time to draft the detailed regulations necessary to comply with it. In the UK, compliance with the Directive is achieved through the *Ionising Radiations Regulations 1985*, drafted under the Health and Safety at Work Act, which came into effect on 1 January 1986.

14.4 REGULATORY FRAMEWORK UNDER THE HSW ACT

The regulatory framework adopted by the HSC consists of a package which comprises the regulations and an approved code of practice entitled *The Protection of Persons against Ionising Radiations from any Work Activity*. This approved code is in three parts:

Part 1 General
Part 2 Application of Regulation 6 (dose limitation) to specified processes
Part 3 Exposure to radon

14.4.1 Regulations

The basic principle in the regulations is that all necessary steps should be taken to reduce so far as is reasonably practicable the extent to which people are exposed to ionizing radiation. This means that it is not sufficient merely to observe the dose limits specified in the regulations; there is a further requirement to weigh the costs of the possible health detriment from exposure against the costs of reducing or eliminating that exposure. Additionally, the regulations specify a dose value, set at three-tenths of the annual dose limits for workers aged 18 years or over, at which an investigation has to be carried out to ensure that doses are being kept as low as reasonably practicable and to initiate remedial measures if they are not.

The dose limits specified in the regulations for various categories of person (workers, trainees, members of the public, women of reproductive capacity) refer to the sum of all radiation absorbed and committed from both external and internal sources, whether to the whole body or part of the body, arising from work activities.

To facilitate the control of doses the regulations require controlled and supervised areas to be identified where there is a likelihood of people receiving doses in excess of three-tenths and one-tenth respectively of the annual dose limit for workers aged 18 years or over. Certain other criteria, such as the instantaneous dose rate, may also require the establishment of controlled or supervised areas. Any person entering a controlled area must be designated a classified person unless he or she enters under a written system of work designed to ensure that a significant dose cannot be received.

Other regulations deal with appointed doctors and with the review of medical findings. There has to be a system for regular assessment of doses received by classified persons and dosimetry services approved by the HSE must be used for this purpose. In addition there are requirements to provide appropriate safety devices, warning signals, handling tools, etc; to leak-test radioactive sources; to provide protective equipment and clothing and to test them; to monitor radiation and contamination levels; to store radioactive substances safely; to design, construct and maintain buildings, fittings and equipment so as to minimize contamination; and to make contingency arrangements for dealing with foreseeable but unintended incidents. Where large quantities of radioactive substances are held there is a requirement to make a survey of potential hazards and prepare a report, a copy of which has to be sent to the HSE.

There are requirements for people to notify the HSE when they use ionizing radiation, when there is a release or loss of radioactive substances, and when someone has received an excessive dose of radiation. Local investigations of excessive doses have to be made and records kept.

The provision of information on potential hazards and the instruction and training of people involved with ionizing radiation are required by the regulations. In addition there are requirements to formulate written local rules covering all radiation protection arrangements and to provide supervision of work involving ionizing radiation; this normally necessitates the appointment of a radiation protection supervisor (RPS). Furthermore, the regulations require radiation protection advisers (RPAs) to be appointed whenever expert advice is needed and specifically whenever an employer has to designate one or more controlled areas.

As regards medical exposures, the doses received by patients undergoing diagnosis or treatment involving the use of ionizing radiation are not taken into account in determining compliance with dose limits. However, doses received by patients as a result of other patients' medical exposures are taken into account. In addition, there is a requirement generally to ensure that the clinical objectives of diagnosis or treatment are achieved with the minimum exposure.

14.4.2 Approved code of practice and other guidance

The Ionising Radiations Regulations contain the fundamental requirements for control of exposure to ionizing radiation. Details of acceptable methods of meeting those requirements are given in the supporting Approved Code of Practice. An approved code of practice does not, in itself, have any legal standing but it would be taken into account by a court when deciding whether or not a person or organization had failed to comply with the regulations under the HSW Act. The code of practice is couched in much more understandable terms than the regulations governing it and it can explain the reason for the procedure which is included and thus aid the reader's understanding.

In addition to the regulations and the approved code of practice the HSE intends to issue detailed guidance on specific items, although this had not been drafted when this edition was compiled.

14.5   SUMMARY OF UK REGULATIONS

Table 14.1 lists the most important regulations dealing with radiological control in the UK.

14.6   BRIEF SUMMARY OF INTERNATIONAL GUIDANCE AND REGULATIONS IN OTHER COUNTRIES

In Section 14.3.4 the EEC Directive on radiological protection was discussed. This Directive lays down the basic standards for the protection of workers and the public against the dangers arising from ionizing radiations within member states. It is binding on member states with regard to the results which have to be achieved. The Euratom treaty also requires member states to make appropriate provisions to ensure compliance with the basic standards and it also requires that provisions on radiological protection be communicated to the Commission in order to achieve the harmonization of such provisions in member states.

Other organizations which undertake work at international level to formulate radiation protection standards are:

The International Commission on Radiological Protection, ICRP (see Chapter 6).
The International Atomic Energy Agency, IAEA, which produces 'Basic Standards for Radiation Protection' (the governments of member states are invited to use these as a basis for formulating national legislation on radiation protection). In addition IAEA has published a number of expert studies in the form of codes of

Table 14.1 Summary of UK regulations

Regulation	Main provisions
Atomic Energy Act (1946)	Promotion and control of atomic energy development
Radioactive Substances Act (1948)	Control of radioactive substances and radiation apparatus
Radioactive Substances Act (1960)	Regulations on keeping and use of radioactive material and the disposal of radioactive waste
Nuclear Installations Act (1965, 1969)	Deals with licensing and insurance of specified sites which have substantial radiological hazards
Radiological Protection Act (1970)	Set up NRPB and Advisory Committee to provide various national services in the radiation protection field
Health and Safety at Work etc. Act (1974)	Set up Health and Safety Commission and Health and Safety Executive to administer regulations concerning the health and safety of persons at work, and of other persons who may be put at risk by the activities of persons at work
The Ionising Radiations Regulations (1985)	The regulatory package consists of the regulations and an approved code of practice, with more detailed guidance notes on certain specific items.

practice, which give practical safety guidelines for specific applications.

The World Health Organization (WHO) has published various reports concerning radiological protection and has also collaborated with IAEA, and other organizations, in producing a number of guides on the protection of the public and workers against nuclear hazards. However, WHO has not issued any regulations in this field.

The International Labour Organization (ILO) adopted, in 1960, a convention concerning the protection of workers against ionizing radiations. In 1963 ILO published a manual of industrial radiation protection which contains a model code of safety regulations concerning ionizing radiations, plus several guides for specific areas.

The Nuclear Energy Agency (NEA) of the OECD issues radiation protection norms from time to time. These norms are based on ICRP recommendations and are revised periodically to take account of the latest ICRP recommendations. In addition, NEA has sponsored or participated in the preparation of a number of studies in the field of radiological safety and collaborates with other international organizations in the establishment of safety standards.

The legislation and regulations dealing with radiological protection in individual countries are quite diverse and complicated, and it is beyond the scope of this book to attempt a discussion of them. Readers who wish to pursue this topic further should consult the bibliography. However, the regulatory framework in four major nuclear countries, France, Germany, Japan and the USA, is briefly outlined below.

France

As a member of the EEC France complies with the requirements of the Directive via a number of decrees and ministerial circulars. The Ministers primarily concerned with radiological protection are:

the Minister for Scientific and Industrial Development, responsible for atomic energy and whose jurisdiction includes the *Commissariat a l'Energie Atomique*;
the Minister for Public Health and Social Security, who is responsible for safeguarding health and ensuring the health and welfare of the population. He controls the specialized technical department, the Central Service for Protection against Ionizing Radiations;
the Minister for the Environment, responsible for classified establishments.

Germany

As it was not clear whether the Basic Law (*Grundgesetz*) covered all aspects of nuclear law, a special act supplementing the basic law was passed in 1959 for this purpose. Most activities relating to atomic energy are dealt with by federal legislation; legislative authority is left with the Länder (the German federal states) only insofar as no federal legislation exists in the field. There is no single central body in Germany in which all the executive responsibilities concerning nuclear energy are vested. Supervisory and licensing powers are allocated either to the federal or land agencies as appropriate. The regulations dealing with the protection of the health and safety of the general public and of persons involved in work having radiation risks are laid down in the First Ordinance on the Protection Against Radiation Hazards and in the Second Ordinance on the Protection Against Hazards caused by Ionizing Radiations in Schools, both made under the Atomic Energy Act of 1959.

Japan

In Japan the control of nuclear energy is quite centralized. In general the competent authorities responsible for licensing nuclear activities and for radiation protection are under the direct authority of the Prime Minister's Office. The Atomic Energy Act of 1955 laid down the main principles governing protection against radiation hazards. Under this act there is a further Prevention Law which is aimed mainly at ensuring public safety by preventing radiation hazards which may arise from radioisotopes and apparatus generating ionizing radiation. Under the direct authority of the Prime Minister's Office are:

the Atomic Energy Commission, which is in charge of preparing, examining and taking the necessary decisions on aspects such as the basic principles of protection against radiation hazards;
the Science and Technology Agency (STA) which provides administrative support for the Atomic Energy Commission;
the Radiation Council, a specialized body which establishes standards concerning radiation protection and levels of radioactivity.

United States of America

In the USA, the Nuclear Regulatory Commission has the prime responsibility for matters concerned with nuclear energy and the use of radioactive materials. The detailed requirements are set out in the

Code of Federal Regulations, Title 10, Energy, known as 10CFR. Part 20 of 10CFR sets out the standards for protection against radiation including permissible doses and levels of radioactivity in effluents.

Under the provisions of the National Environmental Policy Act of 1969, the Environmental Protection Agency has responsibilities in relation to the environmental impact of nuclear energy.

Health physics
laboratory techniques

15.1 BASIC TECHNIQUES

In this chapter some of the basic techniques required in health physics
laboratories are discussed. These include *radioassay*, which is the
measurement of radioactive samples, and the calibration of health
physics instruments.

15.2 RADIOASSAY

15.2.1 Identification of unknown samples

In general, the assessment of a radioactive sample, such as an air
sample filter paper, requires not only measurement of the sample
activity but also identification of the radionuclides present. Indeed, in
most cases precise measurement of the sample activity is not possible
without some knowledge of the nuclide and hence the type and energy
of radiation emitted. The sample must, therefore, be analysed or else
some assumptions must be made about its composition. It is not
usually practicable or necessary to analyse every sample taken and so,
in a particular area, occasional samples are analysed and the results
are applied to other samples taken under similar conditions, perhaps
with the introduction of a safety factor.

The characteristics of a nuclide which it may be possible to
determine to enable its identification are the type and energy of
radiation emitted, and the half-life. The methods by which these
measurements may be made are described in the following sections.

15.2.2 Energy determination

The most convenient method of energy determination is γ-spectro-
metry, which was described in Chapter 7. In the cases of many

commonly encountered radionuclides the observed spectrum enables rapid identification and measurement of sample activity. Similar techniques may be used for α- and β-emitters but the equipment is less readily available.

If the nuclide is a pure β-emitter or if the equipment available does not include a γ-spectrometer, reliance is often placed on β-absorption methods. These involve counting a sample in a β-counting system (e.g. a G-M castle) and observing the count rate with various thicknesses of aluminium placed between the sample and the detector. An absorption graph is obtained by plotting the count rate against the absorber thickness, usually on *log-linear* graph paper (see Fig. 15.1).

The basis of this method is that low energy β-radiation is more easily absorbed than high energy radiation and so, by comparing the absorption curve with curves for β-radiation of known energy, the β-energy of the sample may be estimated. The most precise method of determining the β-energy is Feather's method. This is described in

Fig. 15.1   Typical β-absorption curve

detail in *Experimental Nucleonics* by B. Brown to which the reader is referred.

A simpler but less precise method is to take a series of counts with increasing absorber thickness until the count rate has been reduced to about one quarter of the initial value. The corrected count rate (see Sections 15.2.4 and 15.2.5) is then plotted against absorber thickness and the thickness which would reduce the count rate to one half of the initial value is read off the bottom scale. This is the *half-value thickness* for that β-energy. An example of the method is illustrated in Fig. 15.2. The thickness of aluminium is expressed in units of grams per square centimetre (g/cm^2). This is obtained by multiplying the thickness of aluminium (cm) by the density (g/cm^3) to obtain g/cm^2.

The relationship between the beta particle maximum energy and half-value thickness is shown in Fig. 15.3. Considering the sample in

Fig. 15.2 Determination of half-value thickness for β-emitter

Fig. 15.3 Relationship between half-value thickness and β-ray maximum energy

Fig. 15.2, the half-value thickness is 0.074 g/cm^2 and so, from Fig. 15.3 the β-energy is found to be about 1.4 MeV.

It should be mentioned that such determinations are not always easy because, in practice, samples often contain more than one nuclide and consequently several β-ray energies may be present. Even assuming that the β-energy is determined, this is probably insufficient positively to identify the radionuclide. A further indication may be obtained, in some cases, by measuring the half-life of the nuclide.

15.2.3   Determination of half-life

The concept of half-life of a radionuclide was introduced in Chapter 2. It will be recalled that the activity of a sample at time t is given by

$$A_t = A_0 e^{-\lambda t}$$

where A_0 is the initial activity and λ is the radioactive decay constant for the nuclide. The half-life $T_{1/2}$ is the time at which A_t is one half of the value of A_0, thus

$$\frac{A_t}{A_0} = 0 \cdot 5 = e^{-\lambda T_{1/2}}$$

$$\therefore \ \log_e 0 \cdot 5 = -\lambda T_{1/2}$$

$$-0 \cdot 693 = -\lambda T_{1/2}$$

$$\therefore \ T_{1/2} = \frac{0 \cdot 693}{\lambda} \text{ or } \lambda = \frac{0 \cdot 693}{T_{1/2}}$$

It will be seen that the radioactive decay law can be written in an alternative form

$$A_t = A_0 e^{-0 \cdot 693 t / T_{1/2}}$$

In cases of nuclides having half-lives of between a few minutes and a few months, the half-life can be determined by taking a series of counts on the sample at suitable intervals (such that the count rate decreases by 10–15 per cent between counts). The corrected count rate is then plotted against time on log-linear graph paper, giving a straight line. The time to reduce the count rate to one half of the initial value can then be read from the graph.

Example 1 Using log-linear graph paper, plot a decay curve from the following data and estimate the half-life of the nuclide:

Time (h)	0	2	4	8	12	18
Count rate (counts/min)	6720	6050	5690	4563	3930	2989

It will be found from this plot that the count rate decreases to one half of the initial value (i.e. to 3360 counts/min) in about 15·4 h. Now the sample used to obtain these data is the same sample as used to obtain the absorption curve in Fig. 15.2 and so it is now known that the nuclide emits β-particles of maximum energy about 1·4 MeV and decays with a half-life of 15·4 h. It should now be possible to identify the nuclide by referring to tabulations of nuclide data in order of half-life or β-particle energy (see, for example, *The Radiochemical Manual*). The nuclide present in the sample was sodium-24 which has a maximum beta particle energy of 1·39 MeV and a half-life of 15·4 h.

As with β-absorption measurements, the procedure is often more difficult in practice because of the presence of more than one nuclide. This is illustrated in Fig. 15.4 which again shows a decay plot for sodium-24 but in this case there is also present a short-lived nuclide. After an initial relatively rapid decay, the rate of decay decreases and eventually gives a straight line. The contribution from the longer-lived nuclide to the initial count rate can be obtained by extrapolating the straight line back to zero time. The half-life of this nuclide can then be obtained as before. To determine the half-life of the short-lived component it is necessary to subtract the extrapolated values of the count rate due to the long-lived nuclide from the observed count rate (i.e. the difference between the curve and the straight line is plotted). This procedure gives the decay plot for the short-lived nuclide and the half-life can then be determined. In Fig. 15.4 the short-lived nuclide is chlorine-38 which has a half-life of 37·3 min.

15.2.4 Determination of sample activity

The conditions of measurement used in the radioassay of a particular

Fig. 15.4   Decay curve of mixed ^{38}Cl and ^{24}Na sample

radioactive sample depend on a variety of parameters, such as:

(a) the activity of the sample,
(b) the half-life of the sample,
(c) the precision required,
(d) the radiations emitted by the sample.

In general it is necessary to use a high sensitivity counting system, comprising a radiation detector, a stabilized EHT supply, an amplifier, a discriminator and a scaler which registers the pulses received. (The basic features of counting systems have been discussed in Chapter 7.) Often the detector has to be mounted in a lead castle to reduce the number of counts arising from the background gamma radiation in the laboratory. The counts registered on the scaler must be corrected for background and may require to be corrected for the resolving time of the apparatus. It may also be necessary to calculate the appropriate statistical error.

In certain cases it is possible to mount a very thin radioactive sample at the centre of the sensitive volume of the detector and count all the emitted particles or photons; this is known as 4π geometry counting. In the more common counting systems only a fraction of the emitted particles or photons can enter the detector. The fraction of particles counted compared with the total number emitted is known as the *efficiency* of the counting system. The efficiency of a counting system is usually determined by placing a standard source in the appropriate counting position and determining the number of counts recorded in a fixed time. This is then divided by the emission rate of the source to obtain the counting efficiency. Usually the efficiency is multiplied by 100 and expressed as a percentage.

Example 2　A standard source of 220 Bq gives an uncorrected count of 2045 counts/min in a G-M castle in which the background count is 65 counts/min. What is the efficiency of the system?

$$\text{Efficiency (\%)} = \frac{\text{corrected counts/min}}{\text{source emission rate, dis/min}} \times 100$$

Corrected count rate $= 2045 - 65 = 1980$ counts/min. Now 220 Bq corresponds to $1{\cdot}32 \times 10^4$ disintegrations per minute (dis/min). Therefore

$$\text{Efficiency} = \frac{1980 \times 100}{1{\cdot}32 \times 10^4} = 15\%$$

Example 3 Calculate the activity of a source which gives an uncorrected count rate of 4925 counts/min in the equipment mentioned in the previous example (Answer: 540 Bq).

Apart from the geometry of the counting system, a number of other factors influence the counting efficiency. These include backscatter, self-absorption in the source, absorption in the counter window, absorption in the air-gap between source and detector, all of which depend to some extent on the energy of the radiation. Ideally, the instrument should be calibrated with a source of the same nuclide and the same geometry as the samples to be counted. In the evaluation of routine samples for health physics purposes, a high degree of accuracy is not normally required and it is usual to calibrate the equipment with one typical source and to use the same calibration for the majority of samples. It is important to be aware that some nuclides, particularly low energy β-emitters, can be seriously underestimated by this procedure.

15.2.5 Corrections for resolving time

When a charged particle or photon produces an interaction in the sensitive volume of a detector, there is a short period (usually of the order of 100 μs) during which further events cannot be recorded. This time is known as the *resolving time* of the apparatus and is the time required for the ions to be collected. In many counting systems it is easier to introduce a fixed *dead time* into the system rather than determine the actual dead time experimentally. Since this artificial dead time is a function of known circuit parameters it has a precisely defined value.

Consider a circuit which has a dead time of 200 μs and suppose that it is used to count a sample which gives an observed counting rate of 500 counts/s. In recording 500 counts in one second the counter has been shut down for 200 μs after each count and was therefore inoperative for 500×200 μs ($= 100\,000$ μs $= 0.1$ s). The 500 counts were therefore recorded in a counting time of only 0.9 s and so the true counting rate is

$$\frac{500}{0.9} = 555 \text{ counts/s}$$

This is written mathematically as

$$C = \frac{c}{1 - ct}$$

where

$$C = \text{true count rate}$$
$$c = \text{observed count rate}$$
$$t = \text{dead time}$$

Care should be taken with the units in this equation; if C and c are in counts/s, t must be in seconds (e.g. 200 $\mu s = 200 \times 10^{-6}$ s).

It is inadvisable for dead time corrections of much greater than 10 per cent to be applied, so when the counting rate is very high it is better to reduce it by changing the geometry.

Example 4   Calculate the true counting rate of a sample if the observed counting rate is 15 000 counts/min and the dead time is 300 μs. (Answer 16 216 counts/min.)

15.2.6   Counting statistics

The radioactive decay of single atoms is random in time and so the number of particles or photons counted in a given time will fluctuate about an average value. The *standard deviation* (σ) is a measure of the scatter of a set of observations about their average value. In a counting system, if the average of a large number of counts is $\bar{N}$, then the standard deviation is found to be $\sqrt{\bar{N}}$. Thus if

$$\bar{N} = 400 \text{ counts}, \ \sigma = \sqrt{400} = 20$$

If a single count is made over a time t and N counts are recorded, the standard deviation on the count may be taken as $\sqrt{N}$. Usually it is the counting rate which is of interest and this may be written

$$\text{counting rate} = \frac{N}{t} \pm \frac{\sqrt{N}}{t}$$

Example 5   A 5 s count on a sample gives a result of 100 counts. What is the counting rate and the standard deviation?

$$\text{Counting rate} = \frac{100}{5} \pm \frac{\sqrt{100}}{5}$$

$$= 20 \pm 2 \text{ counts/s}$$

The significance of standard deviation is that 68 per cent of observations are within one standard deviation of the true counting rate. Hence, in the above example, there is a 68 per cent chance that the counting rate lies between 18 and 22 counts/s. The standard deviation is, therefore, a measure of the accuracy of an observation. In counting, greater accuracy can be achieved only by increasing the total count recorded. Thus, if in the previous example 1000 counts had been recorded in 50 s,

$$\text{counting rate} = \frac{1000}{50} \pm \frac{\sqrt{1000}}{50}$$

$$= 20 \pm \frac{31 \cdot 6}{50}$$

$$= 20 \pm 0 \cdot 63 \text{ counts/s}$$

or if 10 000 counts were recorded in 500 s

$$\text{counting rate} = \frac{10\,000}{500} \pm \frac{\sqrt{10\,000}}{500}$$

$$= 20 \pm \frac{100}{500}$$

$$= 20 \pm 0 \cdot 2 \text{ counts/s}$$

Although the counting rate is the same in all three cases, the accuracy has been improved by counting for longer periods.

Standard deviation can also be expressed as a percentage, for example on counts of 100, 1000 and 10 000:

$$N = 100, \sigma = 10 = 10\%$$
$$N = 1\,000, \sigma = 31 \cdot 6 = 3 \cdot 16\%$$
$$N = 10\,000, \sigma = 100 = 1\%$$

For a count (or a count rate) to be accurate to 1 per cent standard deviation, the counting period must be long enough to give at least 10 000 counts.

Usually we require to find the counting rate S due to a source superimposed on a background counting rate. If N counts are recorded in time t_1, due to source and background, and B counts are recorded in time t_2 due to background alone, then the corrected counting rate S is given by

$$S = \frac{N}{t_1} - \frac{B}{t_2} \pm \sigma_s$$

where

$$\sigma_s = \sqrt{(\sigma_1^2 + \sigma_2^2)}$$

$$= \sqrt{\left(\frac{N}{t_1^2} + \frac{B}{t_2^2}\right)}$$

The general expression for this type of measurement becomes

$$S = \frac{N}{t_1} - \frac{B}{t_2} \pm \sqrt{\left(\frac{N}{t_1^2} + \frac{B}{t_2^2}\right)}$$

Example 6 A source is counted in a G-M castle and registers 6720 counts in 4 min. The background is then counted for 10 min and gives 480 counts. What is the corrected count rate and the standard deviation?

$$S = \frac{6720}{4} - \frac{480}{10} \pm \sqrt{\left(\frac{6720}{16} + \frac{480}{100}\right)}$$

$$= 1680 - 48 \pm \sqrt{(420 + 4 \cdot 8)}$$

$$= 1632 \pm \sqrt{424 \cdot 8}$$

$$= 1632 \pm 20 \cdot 6 \text{ counts/min}$$

When counting very low activity sources, very long counting times may be required to achieve an acceptable statistical accuracy. Under such circumstances it is desirable to choose the most efficient distribution of the time between the (source + background) and background counts. The highest accuracy is achieved when

$$\frac{t_1}{t_2} = \sqrt{k}$$

where t_1 is time spent counting (source + background), t_2 is time spent counting background, and k is the ratio of total counting rate (source + background) to the background rate.

It is often worthwhile, when confronted with a sample of unknown activity, to do a preliminary investigation of the counting rates to be expected. This entails a short (source + background) count, followed by a background count, each count being for one or two minutes duration only. From these results the expected counting rates are determined roughly and the length and distribution of time for the accurate assessment of the sample are calculated.

Example 7   If the total counting rate (source + background) is 360 counts/min and the background counting rate is 40 counts/min what proportion of the total available time should be spent on counting background? (Answer: 25 per cent.)

15.3   CALIBRATION OF RADIATION MONITORS

15.3.1   General

The manufacturer of an instrument is normally responsible for carrying out a detailed calibration procedure before handing it over to the user. The following items are normally investigated:

(*a*) sensitivity of the instrument under normal working conditions,
(*b*) energy response,
(*c*) rate response,
(*d*) temperature variations.

The importance of checking the energy response over a wide range of energies (usually 100 keV to several MeV) has already been mentioned in Chapter 8. Similarly, if the instrument has several scales, as in most dose-rate meters, it must respond satisfactorily on all the scales. Temperature response should not be an important factor with laboratory instruments since it is always possible to select components which are practically unaffected by temperature variations.

The user of an instrument merely requires to know that the instrument is within specification when supplied and will subsequently carry out less extensive tests to check on its performance. Most instruments have built-in checks such as battery checks, zero adjustments and sensitivity checks. Generally, the user need only check the sensitivity periodically since this is the parameter which is most likely to change over a period of time. There are basically two methods available for carrying out such checks:

(*a*) direct calibration using known standard sources,
(*b*) intercalibration with a specially calibrated instrument.

In the UK, the Ionizing Radiations Regulations require that all instruments be tested at intervals not exceeding fourteen months and the results recorded in a special register. Similar requirements apply in other countries.

15.3.2   Direct calibration

When only a few instruments have to be calibrated it is usually sufficient to use a simple jig, as shown in Fig. 15.5. This jig ensures that the instrument and source are always positioned in such a way that constant geometry is maintained. On the other hand, when many instruments have to be calibrated it is often necessary to set up a special calibration facility with free-air conditions (that is, minimum scattering around the source) and an accurately-known source. One such facility is illustrated in Fig. 15.6. The instrument being calibrated is moved along a guiding track and positioned remotely. Readings are obtained via a mirror system.

Fig. 15.5   Simple calibration jig

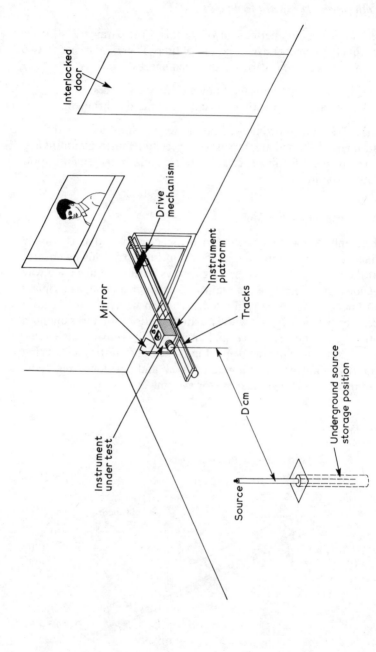

Fig. 15.6 Shielded calibration room with remote operation (after Barnes and Taylor)

Fig. 15.7 Layout for calibration using a standard instrument (after Barnes and Taylor)

15.3.3 Intercalibration with specially-calibrated instrument

When a calibrated instrument is available it is not necessary to have free-air facilities. All that is required is a system which will allow the radiation intensity to be varied over the range required. Such a system is shown in Fig. 15.7, where the radiation intensity is varied by altering the position of the source. The dose rate is measured using the specially-calibrated instrument and then compared with the reading from the instrument under test. It is important to remember that, since no attempts are made to avoid scattered radiation in this test set-up, a variable proportion of the radiation reaching the detector will be degraded in energy. This does not matter when the standard instrument and the one under test have the same energy response. However, if their energy responses are different the results may be invalid.

SUMMARY OF CHAPTER

Radioassay: measurement of radioactive samples.

Identification of nuclides: α-, β- or γ-spectrometry, β-absorption or half-life measurements.

Determination of sample activity: counting efficiency – dependence on energy. Correction for counter background.

Resolving time: detector is inoperative for short time after registering a pulse – reduces effective counting time – important at high count rates.

Counting statistics: standard deviation (σ) – measure of accuracy of a count. $\sigma = \sqrt{N}$. To achieve good accuracy counts must be large – 10 000 counts gives 1 per cent accuracy. Background count – effect on accuracy.

Calibration of instruments: use standard source or compare with calibrated instrument.

REVISION QUESTIONS

1. Why is it usually necessary to determine the composition of the activity on an air sample filter paper? How can the nuclides be identified?

2. A sample gives a count of 16 347 in 1 min when counted in a G-M counter of background 750 counts in 10 min, dead time 300 μs and efficiency 15 per cent.

Calculate the sample activity.

3. What is meant by the resolving time of a detector and what is its importance in counting measurements?

4. The results of β-absorption and decay measurements on a sample are shown below. Estimate the half-life and the maximum β-energy. Refer to nuclide data and attempt to identify the nuclide present in the sample.

β-absorption measurements

Absorber (g/cm^2)	0	0·017	0·041	0·067	0·094	0·120	0·143	0·168
Corrected counts/m	3613	3324	2867	2310	1897	1563	1306	1097

Decay measurements

Time days	0	1	3	6	9	12	15	18
Corrected counts/m	3613	3376	3136	2637	2353	1980	1768	1510

5. What is the significance of the standard deviation of a measurement and how is its value calculated for a single counting measurement?

6. A sample is counted in a β-castle and registers 22 501 counts in 50 min. The background count gives 2040 in 30 min. Calculate the corrected count rate and the standard deviation.

7. It is required to calibrate a gamma dose-rate meter using a standard source which is known to give 720 μSv/h at 1 m. The instrument is to be checked at dose rates of 200, 500, 1000, and 2000 μSv/h. Calculate the distance from source to instrument in each case.

16

Radiological emergencies

16.1 INTRODUCTION

A radiological emergency may be defined as any situation which gives rise to an abnormal or unexpected radiation hazard. This definition covers anything from a minor laboratory spill involving a few MBq of radioactive solution up to a major reactor accident in which many thousands of TBq of fission products may be released. Basically an emergency can arise because of:

(a) a *loss of shielding* – resulting in high radiation levels,
(b) a *loss of containment* – resulting in a release of activity, or
(c) an *uncontrolled criticality*, which is, effectively, the rapid generation of a large radioactive source and high levels of radiation.

Usually, these situations result from some conventional cause such as a mechanical failure, fire, flooding or a transport accident.

It is usual to differentiate between the various levels of emergency which can occur. The minor laboratory spills mentioned above are more of a nuisance than a danger and it is more appropriate to refer to them as *local incidents*. A more serious situation, perhaps requiring evacuation of certain areas, but having no effect outside the site or establishment is often called a *site emergency*. If the incident could be hazardous to the general population outside the site, it is sometimes termed a *public emergency*.

In any event it is of great value to have analysed the possible occurrences in advance and to have formulated procedures for dealing with them. It is of fundamental importance that any abnormal situation is detected as quickly as possible. For example, if a loss of shielding accident is detected immediately and either corrective action or evacuation measures taken, the dose received may be very small. If, on the other hand, operating personnel and others who might be exposed are not aware of the accident, very large doses could be received.

In this chapter a number of situations of the types listed above will be considered.

16.2 LOSS OF SHIELDING

16.2.1 Small sealed source

Small sealed sources, usually γ-emitters, are widely used in industry, medicine and teaching. It is unlikely that a source of less than say 100 MBq could result in excessive dose to a person (unless, for example, it was carried in the pocket) and so loss of shielding would probably be a *local incident*. Such sources are usually handled by tongs and stored in small lead-lined pots. The most common loss of shielding incidents occur when a source is removed from its container by some person who fails to return it. The best method of protecting against this situation is to use installed alarm instruments which show when the source is out of its pot. Regular source musters minimize the possibility that the situation will remain undetected for a long period of time.

The loss of shielding could also result from mechanical damage, for example, if the container was dropped, in which case there should be no problem of detection, or from the effects of fire which might not only cause the shielding to melt but also cause a loss of containment of the source material itself.

16.2.2 Large sealed sources

Large sealed sources such as those used for industrial processing or radiography and medical radiotherapy are usually housed in specially constructed containers with mechanical means for controlling the time of exposure. The containers are designed to withstand normal mechanical accidents and to resist fire. The possibility of inadvertent exposure of the source is minimized by the design of the equipment but alarm systems are desirable to detect fault conditions.

The majority of accidents involving sources of this type have been in industrial radiography. This process is often performed in difficult conditions on construction sites where no form of installed monitoring equipment is used. A number of accidents have occurred in which the radioactive source had become detached from the operating mechanism and when the mechanism was retracted into the storage position the source remained unshielded. In some cases the source was found by some person who, being unaware of its nature, placed it in his pocket. As a result very large, and sometimes fatal, doses were received. Apart from using the appropriate equipment, prevention of this sort of situation requires strict adherence to a monitoring procedure by the radiographer. Various portable alarm devices are also available which can be used in field conditions.

16.2.3   Reactor fuel handling accidents

The problems and dangers involved in handling the intensely radioactive fuel from a nuclear reactor are described in Chapter 10. On the large power generating reactors complex, remotely controlled handling equipment removes fuel from the reactor and transfers it to a cooling pond. It is virtually impossible, because of built in safety devices, for fuel to become unshielded at this stage. On research reactors there is usually more scope for fuel handling accidents and greater reliance has to be placed on the use of approved operating procedures. There is greater scope for loss of shielding accidents in fuel cooling ponds, either because fuel is inadvertently withdrawn from the pond or because of loss of water. The possibility of such occurrences is again minimized by good design of equipment but as a final safeguard an installed radiation alarm system is essential.

16.3   LOSS OF CONTAINMENT

16.3.1   Minor spillage of radioactivity

Perhaps the most common 'abnormal occurrence' is a minor spillage of up to a few MBq of radioactive solution. The frequency of these is minimized by good laboratory practice such as keeping containers of solutions in trays which will contain any spillage. Spills do, however, occur even in the best regulated laboratories, but, if they are dealt with correctly, the contamination and therefore the incident should not spread outside the laboratory or area in which it occurred. After any personnel decontamination which may be required, the most important action is to wipe up the activity using absorbent materials before it dries out and becomes airborne.

A useful precaution in laboratories handling unsealed sources is to have available a few *spillpacks*. A spillpack is simply a plastic bag containing a pair of gloves, a pair of overshoes and a wad of absorbent material (cotton waste, paper towels, etc.). The gloves and overshoes are donned and the spill is wiped up using the absorbent material which is then replaced in the bag for disposal. Alternatively, if any person is contaminated as a result of the spill, by putting on the clean gloves and overshoes he or she is able to go to a change or decontamination area without spreading contamination. Having quickly cleaned up the bulk of the activity the surface can then be monitored and further decontaminated as necessary.

A spillage which remains undetected for some time is more likely to cause problems because it will be spread around, and possibly outside, the area. This is why it is important that people who work in areas in which unsealed sources are used should make a habit of washing and monitoring each time they leave the area.

A release of activity can also result from a failure of services such as ventilation or electrical supplies. Glove boxes can pose a particular problem in this respect. A glove box is normally operated at a pressure slightly below atmospheric which means that leakage tends to be in rather than out. If some failure causes the box to pressurize, leakage occurs outwards or, more seriously, a glove or panel may be blown out resulting in a major release.

16.3.2 Major spills of radioactivity

A major spill involving upwards of 100 MBq or so of activity could result in a serious hazard depending on the radiotoxicity of the nuclides invoived. This would require immediate evacuation of personnel, shutdown of the ventilation system and sealing off the area in order to contain the spread of activity. A controlled re-entry to the area by a team wearing appropriate protective clothing and respirators would be necessary. It is under these circumstances that the value of a properly designed laboratory is apparent. Decontamination is a relatively simple matter in a good laboratory where proper attention has been paid to surface finishes. In badly designed laboratories, decontamination may be difficult or impossible.

16.3.3   Releases of fission products from reactors

The most serious type of loss of containment accident involves fission products from a reactor. It will be recalled that they are contained by the fuel cladding, the boundary of the cooling system and the reactor building. In a power reactor, the most likely cause of a release of fission products is a failure of the pressure circuit (e.g. the fracture of a coolant duct) resulting in a loss of coolant, overheating and melt-down of the fuel. Fission products may then be released from the molten fuel and escape through the breach in the cooling system. If, in a large reactor, only 0·1 per cent of the fission product inventory leaked from the cooling system this could amount to over 10^{17} Bq (3 MCi); if 1 per cent of this amount escaped from the reactor building or *containment*, the release to the environment would be 10^{15} Bq (30 kCi). This would result in very high levels of radiation and contamination on the reactor site. It would also be a hazard to the local population and hence a *public emergency*.

It was considerations such as this which led to the siting of the first generation of nuclear reactors in remote areas. Large plants of more recent design are being sited quite close to towns. This is justified either because in plants such as the advanced gas-cooled reactor (AGR), the entire reactor system is inside a massive pre-stressed concrete pressure vessel and a failure of this is virtually inconceivable, or because other reactor systems such as LWR have special systems, including containment structures, to minimize the release of radioactivity in the event of a failure of the coolant system.

One of the earliest reactor accidents occured at Windscale, Cumberland in 1957. The reactor was of a very early design and used direct cycle air cooling, i.e. air was drawn in through the reactor core, removing heat, and was discharged back to the atmosphere through filters and a tall stack. A special operation was being performed which caused the fuel rods to overheat and catch fire. The main activity released was iodine-131 which being a vapour was not removed very effectively by the filters. An estimated 7×10^{14} Bq (~ 20 kCi) of ^{131}I was released and, although evacuation of the local population was not necessary, milk produced in a large area downwind of the site was declared unfit for consumption. This was because of the exposure pathway: ^{131}I – pasture – uptake by cows – milk – consumption of milk – dose to thyroid. Other fission products such as strontium and caesium were also released but, because of the action of the filters, in much smaller quantities.

Direct cycle air cooling is no longer used on power reactors and so further accidents of this type are not possible. A great deal was learned about the sort of organization, equipment and procedures which are necessary to deal with major accidents.

In 1979, an accident occurred at a large commercial PWR plant at Three Mile Island (TMI), Pennsylvania, USA. This arose from a major leak in the pressure system and the failure of a safety system to operate because of an incorrectly aligned valve. The resulting loss of cooling led to overheating of fuel and release of activity to the environment via the gas waste system. The activity released was mainly the short-lived inert fission product gases, and the resulting radiation exposure of the surrounding population very low. However, some evacuation was undertaken, mainly of young children and pregnant women, and the accident had a considerable adverse effect on public attitudes to nuclear power.

The TMI accident has led to increased efforts to understand the processes which would occur during a severe reactor accident. It now seems that for most types of accident which could occur on light water reactors the release of fission products would be considerably less than had been thought. This is because various processes within the containment, and the containment structure itself are very effective in limiting the release.

The most serious reactor accident ever to occur was at Chernobyl, USSR, in 1986 and involved a 1000 MWe graphite moderated boiling water reactor – a design peculiar to the USSR. At the time of writing the details of the accident are not known but it appears that at some stage failures occurred in pressure tubes which separate the red hot graphite from the cooling water. This led to a massive explosion and fire which devastated the reactor and its containment and resulted in a very large release of radioactivity. Two people were killed and some tens of people suffered acute radiation injuries. About 50 000 people were evacuated from the downwind sector and significant air and ground contamination occurred some hundreds of kilometres from the site.

16.4 UNCONTROLLED CRITICALITY

16.4.1 General

The process of fission and the conditions under which a chain reaction can occur are described in Chapter 10. Uncontrolled critical

excursions could occur in any plant or laboratory in which sufficiently large quantities of fissile materials are handled, and in reactors. The main feature of an uncontrolled criticality is the intense
prompt neutron and gamma radiation given off during the excursion.
If it occurs in an area where there is little or no shielding a very large
external hazard results. However, if it occurs in the core of a reactor,
the hazard would be greatly reduced by the biological shield. In either
situation, if the energy released is large enough it can result in an
explosive reaction, loss of containment and an internal hazard.

There are three approaches to the prevention of criticality when
large quantities of fissile material are present, these are: (*a*) provision
of *neutron absorbers*, (*b*) the use of *safe geometry*, and (*c*) limitation of
quantity (*batching*). In a reactor, method (*a*) is normally used to a
large extent but in fuel plants methods (*b*) and (*c*) are used either
separately or in combination.

16.4.2   Reactor fuel plants

There are three types of reactor fuel plant, namely, enrichment, fuel
fabrication and irradiated fuel reprocessing plants, all of which
handle large quantities of fissile material. The material may be in solid
form or in solution, the latter form being more hazardous because of
the neutron moderation provided by the solvent.

The safe geometry method involves making all process vessels,
tanks and pipework of such a shape that their contents cannot go
critical. The most efficient shape to produce a critical arrangement is a
sphere since in this configuration neutrons are least likely to escape
without causing further fission. Conversely the safest shapes are thin
slabs or tall cylinders.

The safe geometry method can also be applied to the handling of
fissile materials in solid forms such as billets, rods or plates of fuel. An
example of this is the *thin-layer method*. Here, the essential feature is
that within a given area all fissile material is stored, processed,
transported and generally handled within a certain layer. For example, if the safe thickness for the type of material being handled is
0·15 metre, all fissile material would be stored, processed, etc., at a
specified height, say between 1·0 metre and 1·15 metre above floor
level. All working surfaces would be 1·0 metre high, and trolleys,
machines and storage racks would be arranged so that the material
always remained in the thin layer.

Batching means that the fissile material is processed through the plant in quantities which are too small to go critical even under the worst geometry. In order to provide a good margin of safety, batches are usually small enough to ensure safety even if double-batching occurred because of a mechanical or administrative failure. Another important point is that vessels and batches must be adequately spaced to prevent interaction between them.

Whichever method of criticality control is used allowance must be made for contingencies. In particular the possibility of flooding must be considered because of the moderation and reflection provided by water. Plans for fire-fighting are often complicated by the need to preclude water from the area.

Fuel plant accidents are typified by that which occurred at Los Alamos, USA, in 1958. Whilst an inventory was being taken of plutonium residues, the contents of two tanks were drained into a third tank. The two tanks had contained a safe quantity but when added they constituted an unsafe quantity. The residues were fairly heavy and settled in the bottom of the tank in a subcritical configuration. However, when the tank was stirred (electrically) the residues mixed with the solvent which provided neutron moderation and the system went critical. The operator received a fatal dose estimated at 120 Gy (12 000 rad).

16.4.3 Reactors

The majority of uncontrolled critical excursions which have occurred on reactors have involved low power experimental facilities, rather than large power reactors. The accidents were, in most cases, due to a combination of circumstances such as a feature of bad design, a mechanical or electrical failure and an operator error.

The most comprehensively investigated and reported accident was on the SL1 reactor at Idaho Falls, USA, in 1961. Following a routine shutdown for maintenance, an operating crew of three men was reassembling the control rod drives in preparation for startup. The design of the rod drive mechanisms was such that the rods had to be raised manually a few cm whilst they were being connected. It appears that the central control rod was manually withdrawn about 0·5 metre causing the reactor to go critical. The energy released caused a violent steam explosion which killed the three operators. Recovery operations were hampered by radiation levels of about 10 Sv/h

(1000 rem/h) inside the reactor building due to fission products released from the core. Very little release of radioactivity from the building occurred even though it had not been designed as a containment. The accident resulted from a serious design fault and inadequate supervision or training of operators.

16.5   PRE-PLANNING FOR EMERGENCIES

One important result of the TMI accident, mentioned previously in Section 16.3.3, was that all countries reviewed their pre-planned arrangements for nuclear emergencies. In the USA this resulted, amongst other things, in an increase in the distance from the plant for which detailed pre-planning was required. In the UK the review group concluded that the distances did not need to be increased but called for a number of actions to improve the coordination of emergency services and to streamline the decision-making process in the event of an emergency.

Pre-planning to deal with emergency situations begins at the design stage of any plant, process or experiment. A detailed safety analysis at this stage not only indicates the major hazards but may also enable methods of reducing them to be incorporated in the design. No matter how good the design is, or how many safeguards are provided, there always remains the possibility of an accident and in order to deal with this eventuality an *emergency organization* is required.

The size of the organization depends very much on the type of plant and the possible scale of the emergency. In a large plant such as a power reactor the organization is quite large and includes representatives from different departments. For example:

> *Administration department* can assist with such matters as transport and liaison with external authorities, services, etc.
>
> *Engineering department* is responsible for providing rescue and damage control teams, decontamination services and maintaining emergency equipment.
>
> *Medical department* deals with casualties, radiation or otherwise, and liaises with hospitals and medical authorities.
>
> *Health physics department* provides monitoring equipment and services and advises on all aspects of radiation protection.

The actions required and the responsibilities of the various parties in the organization are detailed in the *emergency procedures*. This document includes instructions on evacuation, monitoring, communications, re-entry and use of *emergency equipment*.

The emergency equipment includes rescue equipment, medical equipment, protective clothing, breathing apparatus and monitoring instruments. In the latter case, it must be borne in mind that very high radiation and contamination levels may occur and so special high range instruments are required. In the SL1 accident, mentioned earlier, the radiation levels encountered by the rescue team were greater than the maximum range of their instruments (5 Gy/h, 500 rad/h). Instruments extending up to about 50 Gy/h (5000 rad/h) are now available for emergency use.

Finally, the importance of exercising the emergency organization is emphasized. No matter what the scale of a potential situation, regular exercises remind staff of their actions and responsibilities, test the emergency equipment and highlight shortcomings in the procedures.

SUMMARY OF CHAPTER

Radiological emergency: abnormal or unexpected hazard.
Various levels of severity: local incident, site emergency or public emergency.
Due to: loss of shielding, loss of containment or criticality – usually as a result of conventional failure.
Detection of situation: vital that an incident is recognized immediately – installed instruments.
Pre-planning: begins at design stage – detailed safety analysis.
Emergency organization: emergency procedures and equipment. Emergency exercises.

REVISION QUESTIONS

1. What is a radiological emergency and how could such a situation arise?

2. Discuss the importance of rapid detection of the situation and explain how this might be achieved in practice.

3. With particular reference to reactor fuel plants, describe the methods by which the criticality risk is controlled.

4. Write a short set of emergency instructions to apply in the event of a spillage in a small laboratory handling about 100 MBq of a low toxicity nuclide.

The organization and administration of health physics services

17.1 THE OVERALL PROCESS

The previous chapters of the book have been devoted to explaining the technical concepts and practice of radiation protection. In this final chapter the more general aspects of organization and administrative control are considered.

The process of radiation protection depends on:

(a) The establishment of standards for radiation protection.
(b) The formulation of regulations and codes of practice to meet the standards.
(c) The design and operation of plants or facilities in accordance with the codes.
(d) The supervision and continuous review of the whole process.

17.2 STANDARDS, REGULATIONS AND CODES

It has been mentioned in earlier chapters that the radiation protection standards used in most countries are based on the recommendations of ICRP. These recommendations are continually reviewed in the light of new evidence from research institutions in many parts of the world. As explained in Chapter 14, countries which are members of the European Communities are subject to the Euratom Treaty and must comply with the standards laid down in the EEC Directives on radiation protection. The EEC Directives reflect the thinking and numerical values of the most recent ICRP recommendations.

In the UK a comprehensive regulatory framework has been established under the Health and Safety at Work (HSW) Act to revise, harmonize and bring up-to-date the existing large body of statutory requirements and voluntary standards on radiological protection, taking into account the EEC Directive. A multi-tiered approach is proposed which will comprise:

(a) *regulations,* which specify duties and requirements, dose limits, record keeping requirements, and so on;

(b) *an approved code of practice,* which explains the detailed application of the regulations in specific areas of work activity and gives guidance on matters of general concern to those involved with ionizing radiations. This code of practice is approved by the Health and Safety Executive (HSE);

(c) *guidance notes,* published under the auspices of HSE, which give more detailed advice and guidance on certain topics, taking into account local circumstances. At the plant level, there are usually detailed procedure documents which interpret the requirements of the official regulations in the context of the particular operations being undertaken.

17.3 DESIGN AND OPERATION

Radiation protection in any plant begins at the planning and design stage. This means that the designer must be familiar with the general concepts of radiation hazards and the means of controlling them. In addition there must be extensive consultation between the designer and various specialists. These could include chemists, physicists, health physicists, fire prevention officers and industrial safety experts.

The need for advance planning applies not only to the design of the plant but also to the methods of operation and maintenance. As a general principle, where an error in operation could result in a serious radiation hazard, engineered safeguards should be provided either to prevent the error or to limit its consequences. For example, on a reactor, the consequences of an operator error are limited by the reactor protection system which would shut down the reactor if the power or temperature exceeded some pre-determined value.

The problems of maintenance must also be given careful consideration at the design stage. The sort of situation which should be avoided is the location of equipment which needs frequent maintenance in areas of high radiation. Similarly the possibility of releases of radioactive contamination as a result of maintenance operations and the implications for plant layout should be considered. In large and complex plants such as reactors or chemical plants, operation and maintenance is usually in accordance with detailed written procedures.

17.4 THE HEALTH PHYSICS ORGANIZATION

A Health Physics Department is strictly a service organization. Its function is to advise on all matters relating to radiation safety and to provide personnel and equipment to ensure that safety standards are being met. Its attitude should be constructive in that its staff should try to anticipate problems and suggest alternative approaches rather than waiting for problems to arise and then 'stopping the job'. This pre-supposes good communications at all levels with other departments.

In general, members of a health physics organization should be independent and not have other duties in which a conflict of interests could occur. This is not always possible in small organizations where health physics duties may be a part-time responsibility.

In the regulatory framework under the HSW Act a *radiation protection adviser* (RPA) should be appointed by employers whose operations require the designation of *controlled*, or *supervised* areas. (See Chapter 14.) The RPA must have suitable qualifications and experience, which must be notified to the HSE. In addition to the RPA a *radiation protection supervisor* (RPS) should be appointed to provide local supervision. The RPS should be suitably qualified and should have a sound working knowledge of radiological protection in relation to the particular activities for which he is appointed. More than one RPA and RPS may be appointed if the range of the duties or the processes involved warrant it.

In addition to the RPA and the RPS, the health physics organization of a large establishment will usually contain technical support staff, health physics foremen, health physics monitors and clerical staff.

The basis of an efficient health physics organization is a sound programme of routine work. This includes:

(a) The administration of a personnel monitoring service and the keeping of up-to-date dose records.
(b) The performance and recording of routine radiation and contamination surveys in and around controlled areas, and the analysis of the results to observe trends.
(c) In collaboration with the Instrument Department, the provision, maintenance and calibration of monitoring equipment.
(d) The mustering, testing and recording of all radioactive sources.
(e) Provision and regular testing of emergency equipment.

There may be a number of other duties depending on the organization of the establishment as a whole, for example:

(*f*) Provision of protective clothing and equipment.
(*g*) Control of radioactive waste and the keeping of records.
(*h*) Making arrangements for the medical examinations of classified workers.

Then there are, of course, the day-to-day operational aspects which provide much of the interest. These involve the provision of advice and monitoring during special or non-routine operations in which significant radiological hazards could arise.

17.5 DOCUMENTS AND REPORTS

Various records and reports are required to be kept either because of statutory obligation or in accordance with a code of practice.

The *Health Record* is a record of all medical examinations performed on an employee whilst he is a category A worker. The records for a particular person are required to be preserved for 50 years after the last entry.

The *Source Record* contains information on all radioactive sources and the dates and results of all leakage tests. Retention period, 2 years after the last entry.

The *Instrument Record* is used to record details and results of tests of all instruments used for health physics purposes. Retention period, 2 years after last entry.

Radiation dose records are required to be kept in respect of all classified workers. Any worker is entitled to inspect his or her own dose records. The records are required to be retained for 50 years after the last entry. Under the HSE regulatory framework the approved laboratory retains all the records of the assessments that it carries out. Each record includes the worker's national insurance number and the worker has the right to examine his or her dose record. When a worker changes jobs, he or she will be given a termination record detailing the current dose status (purely for information), and the new employer will seek such a record from his or her own approved laboratory which will carry out the necessary consultation.

Survey Records are usually retained for at least 2 years.

Reports of unusual occurrences. All unusual occurrences which result in or could have resulted in abnormal radiation dose should be fully investigated. The reports would normally be preserved for at least 50 years.

Waste disposal records are preserved indefinitely, in particular those records showing the position of burial of solid waste.

17.6 TRAINING

The atomic energy industry has an enviable safety record. This is due to a large extent to the enlightened attitude of the industry to staff training. The legislation and codes of practice require that all persons exposed to radiation in the course of their work be given training in the hazards and the means of controlling them. This may vary from a short talk on the function of the film badge and an outline of local rules, up to a detailed course in health physics, depending on the nature of the plant and the duties of those involved.

For health physics personnel, specialist courses are necessary. Large establishments often run their own training courses for health physics monitors and technicians, whilst in other cases colleges of technology run courses.

In the UK, the accepted standards for this grade are the examinations in *Radiation Safety Practice* of the City and Guilds Institute. For professional health physics staff there are a number of postgraduate courses run by various universities and polytechnics. The National Radiological Protection Board organizes a number of courses on radiological protection, ranging from short familiarization courses to advanced courses for specialist health physicists.

A most important aspect of training for all health physics personnel is plant familiarization. That is, instruction in the processes and engineering aspects of the plant on which they are working. An understanding of the plant is essential if proper advice is to be given and a constructive attitude maintained.

SUMMARY OF CHAPTER

Considerations in routine radiation protection: standards, regulations and codes, design and operation of plant, health physics organization, documentation and reporting.

Health Record: record of all medical examinations while worker is a classified worker.

Source Record: contains information on all radioactive sources with dates and results of all leakage tests.

Instrument Record: record of tests on all health physics instruments, retained 2 years after last entry.

Radiation dose records: for classified workers, retention period 50 years after last entry.

Survey Records: usually retained for at least 2 years.

Reports of unusual occurrences: retained for at least 50 years.

Waste disposal records: preserved indefinitely.

Training: radiation protection; plant familiarization.

Appendix 1

Relationship of units

Radiological quantity	Old unit	SI Unit	Relationship between units	
Activity of a radioactive material	The curie $1\ Ci = 3\cdot7 \times 10^{10}$ dis/s	The becquerel $1\ Bq = 1$ dis/s $10^{3}\ Bq = 1$ kilobecquerel (kBq) $10^{6}\ Bq = 1$ megabecquerel (MBq) $10^{9}\ Bq = 1$ gigabecquerel (GBq) $10^{12}\ Bq = 1$ terabecquerel (TBq) $10^{15}\ Bq = 1$ petabecquerel (PBq) $10^{18}\ Bq = 1$ exabecquerel (EBq)	$1\ Bq = 2\cdot7 \times 10^{-11}\ Ci$ $1\ kBq = 2\cdot7 \times 10^{-8}\ Ci$ $1\ MBq = 2\cdot7 \times 10^{-5}\ Ci$ $\quad = 27\ \mu Ci$ $1\ GBq = 27\ mCi$ $1\ TBq = 27\ Ci$ $1\ PBq = 27\ kCi$ $1\ EBq = 27\ MCi$	$1\ \mu Ci = 37\ kBq$ $1\ mCi = 37\ MBq$ $1\ Ci\ \ = 37\ GBq$ $10^{3}\ Ci = 37\ TBq$ $10^{6}\ Ci = 37\ PBq$ $10^{9}\ Ci = 37\ EBq$
Absorbed dose	The rad $1\ rad = 0\cdot01$ J/kg	The gray $1\ Gy = 1$ J/kg $1\ Gy = 10^{3}\ mGy = 10^{6}\ \mu Gy$	$1\ \mu Gy = 0\cdot1$ mrad $1\ mGy = 100$ mrad $1\ Gy\ \ = 100$ rad	$1\ mrad = 10\ \mu Gy$ $1\ rad = 10\ mGy$ $100\ rad = 1\ Gy$
Dose equivalent	The rem $1\ rem = 1\ rad \times Q$ Q is the quality factor	The sievert $1\ Sv = 1\ Gy \times Q \times N$ $1\ Sv = 10^{3}\ mSv = 10^{6}\ \mu Sv$ N is the product of all other modifying factors (currently taken as 1 by ICRP)	$1\ \mu Sv = 0\cdot1$ mrem $1\ mSv = 100$ mrem $1\ Sv\ \ = 100$ rem	$1\ mrem = 10\ \mu Sv$ $1\ rem = 10\ mSv$ $100\ rem = 1\ Sv$

Answers to numerical questions

Chapter 1

2. 27, 27, 33
3. 1, 1/1840, 1; $+1$, -1, 0

Chapter 2

2. (a) $^{234}_{90}$Th   (b) $^{3}_{2}$He   (c) $^{62}_{28}$Ni
4. 2.3 minutes
5. (a) 5 MBq   (b) 0·75 MBq   (c) 1300 MBq   (d) 1 MBq
6. (a) $3·7 \times 10^4$ (b) $1·85 \times 10^8$ (c) 10^{11} (d) $3·7 \times 10^{16}$ Bq

Chapter 3

6. $2·65 \times 10^7$ n/(m^2 s)
7. 1 mGy, 50 mGy, 1 μSv, 10 mSv
8. $0·84 \times 10^{-3}$ Sv (0·84 mSv)

Chapter 5

2. 1·88 mSv/year $\times$ 30 years $= 56.4$ mSv (5·64 rem)

Chapter 6

3. 500 mSv, 500 mSv, 150 mSv
5. 40 mSv
6. (a) 333 mSv, 417 mSv, 417 mSv, 1667 mSv
   (b) 500 mSv for all four organs
   (c) 333 mSv, 417 mSv, 417 mSv, 500 mSv

Chapter 8

2. 10 h, 25 μSv/h
3. 225 μSv/h, 3 m
4. 6 m
5. 200 μSv/h

Chapter 9

2. Inhalation ALIs: 540, 54, 162, 1.35×10^{-2} (dioxide),
   5.4×10^{-3} (other compounds)
   Ingestion ALIs: 270, 27, 108, 54 (oxide and hydroxides),
   5.4 (other compounds)
3. 12.5 mSv
4. 2×10^5 Bq
5. 8×10^{-2} Bq/m³
7. 4×10^4 Bq/m²
8. 1.67×10^4 Bq/m³

Chapter 15

2. 1970 Bq
4. Maximum β-energy 1.71 MeV, half-life 14.3 days,
   phosphorus-32
6. 382 ± 3.4 counts/min
7. 1.9, 1.2, 0.85 and 0.6 m

Bibliography

GENERAL REFERENCE

Principles of Radiation Protection. Morgan, K. Z. and Turner, J. E. eds., Wiley, 1967.
This is a comprehensive and detailed book aimed at the professional Health Physicist. Each chapter is written by a recognized expert in the subject.

CHAPTERS 1 AND 2

Basic Nuclear Physics. Williams, I. R. and Williams, M. W., George Newnes, 1962.
This book covers all the background nuclear physics that would be needed by a Health Physicist. As far as possible, the authors have adopted a non-mathematical approach to the basic concepts.

CHAPTER 3

Radiation Dosimetry. Volume I. Attix, F. H., Roesch, W. C. and Tochilin, E., eds., Academic Press, 1967.
This is an advanced work on radiation units and dosimetry. It is generally accepted as the standard work in this field.

CHAPTER 4

Atomic Radiation and Life. Alexander, Peter, Penguin Books. 2nd edition, 1965. A thorough account of the effects of radiation on biological systems. It does not assume any technical knowledge and can be understood by readers without a mathematical background.
Nuclear Radiation: Risks and Benefits. Pochin, Edward, Clarendon Press, Oxford, 1983.
UNSCEAR. Ionising radiation: sources and biological effects. 1982 report to the General Assembly, with annexes. New York. 1982 (United Nations Publication E.82. IX.8.)

An annex to this report reviews in detail the available information on radiation-induced cancer, the genetic effects of radiation and the effects of irradiation of the uterus.

ICRU Report No. 30. Quantitative concepts and dosimetry in radiobiology. International Commission on Radiation Units and Measurements. Washington D.C. 1979.

CHAPTER 5

Environmental Radioactivity. Eisenbud, M., McGraw Hill, 1963. A systematic account is given of the behaviour of radioactive substances in the environment. Topics covered include physical and biological transport mechanisms, sources of environmental radioactivity and experience.

The radiation exposure of the UK population – 1984 review. Hughes, J. S. and Roberts, G. C. National Radiological Protection Board report NRPB R173. NRPB, Chilton, Oxon. 1984.

See also the book by Sir Edward Pochin and the UNSCEAR report mentioned above.

CHAPTER 6

The publications of the International Commission on Radiological Protection, Pergamon Press.

The ICRP has issued a number of publications which include both recommendations and reports prepared by specific task groups. Of particular importance are the following publications:

Publication 23 Reference man: anatomical, physiological and metabolic characteristics.

Publication 26 Recommendations of the ICRP.

Publication 30 Limits of intakes of radionuclides by workers.

CHAPTER 7

Radiation Dosimetry. Volume II. Attix, F. H. and Roesch, W. C., eds., Academic Press, 1966.

A comprehensive and detailed review of radiation detection methods by experts in the various fields. The methods reviewed include ionization, chemical, solid state and photographic dosimetry.

CHAPTERS 8 AND 9

Basic Safety Standards for Radiation Protection. Safety Series No. 9. International Atomic Energy Agency, Vienna, 1982.

A Basic Toxicity Classification for Radionuclides. Technical Reports Series No. 15. International Atomic Energy Agency, Vienna.
The IAEA has issued a number of manuals on various aspects of atomic energy and the handling of radioactive substances. Each manual contains a comprehensive discussion of the problems and the recommended methods for dealing with them. These manuals provide very useful ground rules for the Health Physicist.
Design of and equipment for Hot Laboratories, IAEA *Proceedings Series STI/PUB/436,* 1976.
This book describes the requirements for containment, shielding, ventilation, fire protection, criticality control, waste management and safety in operations in radioactive facilities. It includes contributions on the latest developments in these areas, emphasis being placed on the safety features of planning and design.
ICRP Publication 36 General principles of monitoring for radiation protection of workers.
See also the ICRP Publications listed under Chapter 6.

CHAPTER 10

Nuclear Reactor Engineering. Glasstone, S. and Sesonske, A. Van Nostrand, 1963.
This is a sound textbook which covers the fundamental scientific and engineering principles of nuclear reactor systems. The aspects considered in detail include nuclear reactions, reactor theory, control of reactors, reactor materials and fuels, radiation protection, reactor safeguards and shielding.
Engineering Compendium on Radiation Shielding. Jaeger, R. G., ed., Springer-Verlag, 1968.

CHAPTER 11

Objectives, Concepts and Strategies for the Management of Radioactive Waste Arising from Nuclear Power Programmes. A report by a group of experts of the OECD Nuclear Energy Agency, 1977.
The proceedings of a number of conferences held by the International Atomic Energy Agency, including IAEA STI/PUB/433, STI/PUB/406 and STI/PUB/264.
These publications contain the full text and discussion of papers presented by specialists from many countries.

CHAPTERS 12 AND 13

X-rays. Stern, B. E. and Lewis, D. Pitman, 1970.

This is a well balanced and comprehensive account of the production, measurement and applications of X-rays. It is highly recommended to readers who are closely involved with X-rays and their uses.
Basic science of nuclear medicine. Parker, R. P., Smith, P. H. S. and Taylor, D. M. Churchill Livingstone, Edinburgh, 1978.

CHAPTER 14

Nuclear legislation. An analytical study of regulations governing nuclear installations and radiation protection. OECD Nuclear Energy Agency, 1972. A further study by the OECD Nuclear Energy Agency was published in 1980, entitled *Description of Licensing Systems and Inspection of Nuclear Installations.*

CHAPTER 15

Experimental Nucleonics. Brown, B. Iliffe Books, 1963.
The first half of this book is an account of the general principles of radiation detection and measurement. This is followed by detailed notes on a series of experiments suitable for a course in radiation protection.

CHAPTER 16

ICRP Publication 28 The principles and general procedures for handling emergency and accidental exposures of workers, 1978.
Handling of radiation accidents, IAEA *Proceedings Series STI/PUB/463,* 1977. Contributions are included on radiation accident case histories, organization and planning, training and exercises, aspects of the detection and assessment of accident situations and the analysis of accidents.

Index